CHARACTERISATION OF HIGH-TEMPERATURE
MATERIALS: 6
Series Editor: M McLEAN

SURFACE STABILITY

Oxidation - Corrosion - Erosion - Wear

*Proceedings of the sixth
seminar in a series of seven
sponsored and organized by the
Materials Science, Materials Engineering and
Continuing Education Committees of
The Institute of Metals: held in
London on 6 September 1989*

EDITED BY T N RHYS–JONES

THE INSTITUTE OF METALS

1989

Book Number 448

Published in 1989 by The Institute of Metals
1 Carlton House Terrace, London SW1Y 5DB

and
The Institute of Metals
North American Publications Center
Old Post Road, Brookfield VT 05036
U S A

Compiledfrom typematter and illustrations
provided by the authors

Printed in Great Britain

I S B N 0-901462-61-6

CONTENTS

PREFACE

This is the sixth of a series of seven
monographs presenting a practical overview
of current methods of characterising high
temperature materials. To date the series
has covered microstructural, chemical,
physical and elastic characterisation,
mechanical testing and non-destructive
examination. The final monograph will
cover numerical techniques. Although the
series is primarily concerned with high
temperature materials characterisation,
many of the evaluation techniques are
applicable to materials used at any
temperature.

This volume commences with overviews
highlighting the surface stability problems
encountered by materials used in the gas
turbine section of the modern aero engine
and the range of materials used in land
based power generation equipment. Methods
of surface protection and corrosion control
are outlined. The importance of the need
for extensive materials evaluation prior
to service use are highlighted with
particular reference to the selection of
suitable test methods and the elucidation
of degradation mechanisms. The relative
merits of various test techniques used
to evaluate the oxidation, corrosion,
erosion and wear characteristics of high

temperature materials will be described. The importance of understanding the oxidation behaviour of materials will be highlighted in a chapter on the use of the 'oxide dating' technique used by the CEGB to investigate problems and failures in power plant. Finally the techniques available for the chemical and structural analysis of materials and their degradation products will be described with reference to the behaviour of high temperature materials.

The aim throughout is to permit the newcomer or non-specialist to appreciate the range of surface stability problems encountered by high temperature materials, to gain an appreciation of the various techniques used to characterise their oxidation, corrosion, erosion and wear behaviour and to become acquainted with the analytical methods generally used to characterise degraded materials.

The authors have each provided extensive lists of references in order to allow readers to explore the various subjects in more detail as and when required.

Dr. Thomas N. Rhys-Jones
Materials Laboratory
Rolls-Royce plc
P.O. Box 3
Filton
Bristol BS12 7QE
England

1: Surface stability of materials used in aero engine applications
T N RHYS - JONES

Dr T.N Rhys-Jones is the Materials Laboratory of
the Company Materials and Mechanical Technology
Group of Rolls-Royce plc, Bristol, UK.

1.   SUMMARY

The environmental degradation processes encountered
by hot-section components in gas turbines are
described and mechanisms of oxidation, corrosion and
erosion discussed.

The protection of components via the use of surface
coatings is described, the range of coatings and
their performance discussed and the consequences of
coating degradation outlined with reference to aero
engine applications.

2.   INTRODUCTION

The development of high strength superalloys to
permit increased turbine operating temperatures and
hence higher engine efficiencies has generally been
achieved at the expense of the Cr content of the
alloys, with an increase in the γ' precipitate
(Ni_3Al, $Ni_3(Al,$ $Ti,$ $Nb)$) formers such as Al
and Ti. In addition, solution strengthening
elements such as W and Ta are often used. The
development of gas turbine materials has been
extensively reported in the literature (1-5) and so
will not be discussed in detail in this publication.

The improvements in mechanical properties and temperature capability due to the alloy compositional changes outlined above have been generally accompanied by a marked decrease in the corrosion resistance of the alloys due to the decreased Cr content, although their oxidation performance has been relatively unaffected. Figure 1 shows the influence of alloy Cr content on the corrosion behaviour of a range of superalloys in the temperature range 700°- 1100^OC (6).

3.   DEGRADATION OF HIGH TEMPERATURE MATERIALS USED IN GAS TURBINES

The basic forms of environmental degradation encountered by aerofoil components in aero engines are (7, 8):

(i)   low temperature sulphidation, known as Type II corrosion, associated with high SO_3 the combustion product environment, which occurs over the range 500° - 750^OC

(ii) high temperature oxidation, which occurs at 950^OC and above

(iii) erosion, which may occur at all turbine operating temperatures.

Since there is a temperature variation over the surface of turbine vanes and blades, perhaps ranging from 400° - 1100^OC, more than one form of attack may occur on a single component. The exact temperature at which one form of attack to another occurs has been reported to be dependent upon the time at temperature, the temperature itself and the material itself. The extent of corrosion depends on several factors including temperature, gas composition, deposit composition, contaminant flux rate, the presence of erosive media, time and engine operating conditions (7, 8).

3.1   HIGH TEMPERATURE OXIDATION

OXIDATION BEHAVIOUR OF γ' STRENGTHENED Ni-BASE SUPERALLOYS

Extensive studies of the high temperature oxidation of γ' strengthened Ni-base superalloys have shown that their behaviour is generally similar to the Ni-Cr-Al ternary system (9). During the transient stages of oxidation a whole range of oxides is produced, with later steady state scale formation being dependent upon the Cr and Al content of the alloy and on diffusional processes.

The oxidation behaviour of the Ni-Cr-Al system may be divided into three distinct groups and can be represented by the use of a ternary oxidation diagram which may be superimposed on a ternary phase diagram as shown in Figure 2 (9, 10).

The three oxidation regimes are:

(i)   the formation of an external NiO scale with a subscale comprising Cr_2O_3 and/or Al_2O_3

(ii)   the formation of Cr_2O_3 with a discontinuous underlying Al_2O_3

(iii) the formation of a thin continuous exterior Al_2O_3 scale.

It should be noted that the scale compositions may be far more complex when material compositions approach the 'border lines' on the oxidation diagram (Figure 2).

OXIDATION BEHAVIOUR OF Co-Cr-Al ALLOYS

Studies of the oxidation behaviour of Co-Cr-Al alloys have been reported in some detail (11-13). A ternary oxide map for the Co-Cr-Al system is shown in Figure 3 (13).

The oxidation characteristics of the Co-Cr-Al system can be divided into four district groups (13):

3

(i)   CoO doped with Al and/or Cr

(ii)   duplex scales such as CoO-CoO.Co   (Cr, Al)$_2$O$_4$

(iii)   A complex scale comprising Cr$_2$O$_3$ and Co(Cr, Al)$_2$O$_4$ with internal Al$_2$O$_3$

(iv)   external Al$_2$O$_3$.

As in the case of the Ni-Cr-Al system, the oxides formed on the alloy surface are for more complex when the compositional boundaries are approached.

3.2   HIGH TEMPERATURE CORROSION

It is now generally accepted that the corrosion processes occurring in a gas turbine as a result of the disposition of salts and the presence of atmospheric SO$_x$ contamination can be divided into two distinct groups:

BASIC FLUXING

The corrosion of Ni-base systems by Na$_2$SO$_4$ has been studied extensively by Giggins and Pettit (14) and attributed to a process termed 'basic fluxing' which occurs via the following sequence of events:

(i)   an oxygen gradient is developed across the Na$_2$SO$_4$ melt

(ii)   as a consequence of the oxygen gradient, the sulphur activity is increased an metal sulphides are formed at the alloy surface

(iii)   as a result of sulphide formation, the oxide ion content of the Na$_2$SO$_4$ melt is increased and reaches the values where Al$_2$O$_3$ or Cr$_2$O$_3$ in the scale will dissolve in the Na$_2$SO$_4$.

According to Giggins and Pettit (14) and Goebel et al (15) a process is developed whereby SO$_4{}^{2-}$

4

ions diffuse to the alloy surface and as regions of low oxygen activity are attained, these ions sulphidize metals in the alloy, leaving oxide ions which in turn react with the Al_2O_3 or Cr_2O_3 to form products soluble in the Na_2SO_4 melt i.e. aluminate or chromate ions respectively:

$$Al_2O_3 + O^{2-} \rightleftharpoons 2AlO_2^- \qquad \text{----1}$$
(aluminate)

$$Cr_2O_3 + 2O^{2-} + O_2 \rightleftharpoons 2CrO_4^- \qquad \text{----2}$$
(chromate)

(iv) the aluminate and chromate ions formed by the oxide scale - oxide ion reactions diffuse out from the alloy and are precipitated at higher oxygen partial pressures as Al_2O_3 and Cr_2O_3 respectively.

(v) while sufficient sulphate ions are present, the corrosion will proceed at a linear rate; however as the source of sulphate ions is depleted, the attack diminishes and oxygen becomes more plentiful in the corrosion product.

In contrast, Rapp and Goto (16) have reported that the production of oxide ions need not necessarily occur as a result of removal of S from Na_2SO_4. They attribute the increased basicity of the Na_2SO_4 to the reduction process accompanying the oxidation of alloying elements in the degrading alloy:

$$1/2 \ O_2 + 2e^- \rightleftharpoons O^{2-} \qquad \text{----3}$$

In this model, the corrosion process involves dissolution and reprecipitation of oxide with the requirement for this to occur being a negative gradient of oxygen solubility in the Na_2SO_4. It is proposed that the oxide solubility in the Na_2SO_4 decreases as the salt becomes less basic and so reprecipitation of Cr_2O_3 or Al_2O_3 will occur in the Na_2SO_4 near the deposit/gas interface.

Basic fluxing will not proceed indefinitely.

Reising and Krause (17) have reported that as the thickness of a non-protective oxide scale is increased, the Na_2SO_4 is absorbed into the porous scale rather than remaining on the metal or alloy surface, thus greatly diminishing the extent of further attack.

ACID FLUXING

The acid fluxing process differs from basic fluxing in that it is usually self-sustaining and only relatively small amounts of deposits are capable of producing more extensive attack than that associated with basic fluxing (14, 15).

Salt deposits may be acidified by two processes, namely alloy induced acidity and gas induced acidity, the mechanisms being outlined below:-

Alloy Induced Acid Fluxing

The alloy induced hot corrosion process has been reported to proceed via the following sequence of events:

(1) oxides of Mo and W dissolve in the Na_2SO_4 forming molybdates and tungstates respectively, and some SO_3 is displaced from the Na_2SO_4. The dissolution of these oxides into the Na_2SO_4 will depend upon the oxidation characteristics of the alloy. In some alloys, the oxides are formed at the onset on the oxidation processes, whilst in others they are formed at latter stages of oxidation, thus necessitating longer exposure times before WO_3 or MoO_3/salt interactions occur.

(ii) the Na_2SO_4 gradually becomes enriched in these oxides and a point is reached at which Al_2O_3 and Cr_2O_3 will dissolve in the oxide enriched melt as follows:

$$2\ Al + 3W + 6O_2 \rightleftharpoons Al_2O_3 + 3WO_3 \rightleftharpoons$$
$$2\ Al^{3+} + 3WO_4^{2-} \qquad \text{- - - -4}$$

$$2\ Cr + 3W + 6O_2 \rightleftharpoons Cr_2O_3 + 3WO_3 \rightleftharpoons$$
$$2Cr^{3+} + 3WO_4^{2-} \qquad \text{- - - -5}$$

6

$$Co + W + 2O_2 \rightleftharpoons CoO + WO_3 \rightleftharpoons Co^{2+} + WO_4{}^{2-} \quad ----6$$

$$Ni + W + 2O_2 \rightleftharpoons NiO + WO_3 \rightleftharpoons Ni^{2+} + WO_4{}^{2-} \quad ----7$$

These ions then diffuse through the Na_2SO_4 to the outer zone of the melt, where the above reactions proceed in the reverse direction because of the low activity of the refractory metal oxides in this region as a result of their loss to the gas stream. Thus Cr_2O_3 and Al_2O_3 are dissolved at the alloy/melt interface and then reprecipitated as a non-protective scale at the other side of the melt. In addition, the melt is continually enriched in the refractory metal oxide, although the precipitation process results in some of the melt being incorporated in the porous, non-protective oxide scale.

The important features of the alloy induced acidic fluxing process are that a liquid phase is formed immediately above the alloy because of the presence of MoO_3 and/or WO_3 in the Na_2SO_4 and that oxides relied upon for protection (eg. Al_2O_3 and Cr_2O_3) become non-protective as a result of a dissolution -reprecipitation process.

GAS INDUCED ACID FLUXING

Giggins and Pettit (14) and Goebel et al (15) have investigated the influence of gas composition on the Na_2SO_4 - attack of a CoCrAlY material and noted that extensive corrosion occurs at 700°C (i.e. a fairly low temperature) with Na_2SO_4/SO_3, but that the rate of corrosion is drastically decreased when SO_3 is removed from the atmosphere. The attack was found to require the combined presence of Na_2SO_4 and SO_3, since corrosion in SO_3/O_2 alone was negligible.

The corrosion process showed the following interesting points:-

(i) the Na_2SO_4 was distributed throughout the corrosion product

7

(ii) Co diffuses through the corrosion product to the gas interface where oxides and sulphates of this element are formed.

(iii) Cr is converted to Cr_2O_3 close to the corrosion front and little diffusion of Cr is evident.

(iv) Al is preferentially removed from the alloy by the hot corrosion process; it is associated with S and O adjacent to the corrosion front and is present as oxide throughout the corrosion product.

A model was postulated (14, 15) to describe the above form of hot corrosion. At low temperatures, i.e. below the melting point of Na_2SO_4, the salt deposit becomes molten because of $CoSO_4$ dissolution i.e. the formation of a Na_2SO_4 – $CoSO_4$ eutectic mixture. Beneath this molten layer, the alloy begins to react with components from the deposit, with the principal reaction being oxide removal from the melt since oxidation reactions are thermodynamically most favoured. It has been found that SO_3 is more mobile than O_2 in Na_2SO_4 (18) and so it is reasonable to suppose that concentration gradients of both species are developed across the melt. Al is selectively removed from the alloy along with Cr. Co dissolves in the melt, diffuses to the outer zone and is precipated as CoO, which reacts with Na_2SO_4.
The hot corrosion behaviour of Co- and Ni- base materials below the melting point of Na_2SO_4 has also been investigated in detail by Jones and coworkers (19-21). Their results indicate that such hot corrosion proceeds via the processes outlined by Giggins and Pettit (14) i.e. because of the presence of SO_3 and the resulting formation of $CoSO_4$ (or $NiSO_4$); these form molten $CoSO_4$ (or $NiSO_4$) / Na_2SO_4 deposits which are capable of degrading high temperature materials.

At higher temperatures, higher partial pressures of SO_3 are required to form sulphates of Ni and Co. In addition, as the temperature increases, the SO_2/SO_3 equilibrium favours SO_2 rather than SO₃, thereby decreasing the propensity of gas induced fluxing to occur (14, 19-21).

INFLUENCE OF CARBON ON HOT CORROSION

Carbon deposits can accumulate in the combustion chamber of an engine because of incomplete or faulty combustion. The breakaway of these carbon deposits represents a major hazzard to nozzle guide vanes and blades, particularly in the high pressure turbine section.

The deposition of C onto the surfaces of hot section components has been shown to enhance hot corrosion processes as a result of the formation of localized 'reducing' conditions or 'hot spots' with associated low oxygen activities, thus rendering the material more susceptible to other contaminants such as S.

Hot corrosion by Na_2SO_4 has been shown by Goebel and Pettit (22) to be particularly severe in the presence of C, which influences the scale dissolution mechanisms. Under oxidizing conditions in the presence of C deposits, the dissolution of the oxides NiO and Cr_2O_3 was enhanced and penetration of S into the alloy promoted.

McKee and Romeo (23) have postulated that the chemical effect of C on the salt deposit leading to enhanced alloy sulphidation is as follows:-

$$2Na_2SO_4 + 6C + 3O_2 \rightleftharpoons 2Na_2O + 2S + 6CO_2 \qquad \text{----8}$$

Where the S reacts with alloying elements to form metal sulpludes.

INFLUENCE OF CHLORIDE ON HOT CORROSION

It is generally recognised (24, 25) that the hot corrosion of gas turbine materials is enhanced during operation in chloride - containing environments. The influence of chlorides on the hot corrosion of turbine materials has been studied extensively (24, 25) and results indicate that the chloride modifies the initiation stage of hot

9

corrosion by disrupting the integrity of the oxide scales on the materials, consequently rendering them more susceptible to degradation processes. However, the mechanisms via which chloride disrupts the integrity of the oxide scales is yet to be established.

INFLUENCE OF CONTAMINANT QUANTITY

The high temperature corrosion of gas turbine materials has been shown to be markedly influenced by the amount or deposition rate of contaminants and their composition (26-29). Corrosion is also affected by the composition of the surrounding atmosphere (26-29). Evidently, therefore, the extent of corrosion can be decreased by limiting the deposition of salts on hot-section components and by controlling the concentration of atmospheric contaminants such as SO_x, by reducing the S content of the fuel (29).

TOLERABLE OR 'THRESHOLD' LEVELS OF DEPOSITED AND GASEOUS CONTAMINANTS

In view of the dependence of component material degradation on the amount, deposition rate and concentration of corrosive contaminants, it is reasonable to postulate the existence of tolerable or 'threshold' levels of these species with reference to acceptable rates of attack i.e. component lifetimes.

Hefner and Lordi (30) have shown that the life of hot end components in gas turbines can be directly related to the concentration of contaminants in the fuel as shown in Figure 4 for two temperatures. From the figure it is evident that component life is initially governed by oxidation, but once a 'threshold' concentration of impurities (which gives rise to a specific level of salt deposition) is exceeded, the component life is markedly decreased as a result of an increase in corrosion. Rhys-Jones et al (31) have reported that for an industrial gas turbine burning contaminated fuel it is possible to define 'threshold' levels of deposited and gaseous contaminants with reference to

10

preselected acceptable rates of material loss (i.e. component material lifetimes).

EROSION OF TURBINE MATERIALS

Particulates injected into the compressor intake of an aero engine during takeoff or solids generated in the combustion chamber (eg C) may enter the turbine section of the engine and cause erosion of critical gas path components such as nozzle guide vanes and blades (32, 33).

Studies of the behaviour of turbine aerofoil materials under erosive conditions (32, 33, 34) indicate that the following parameters are the principal factors in determining their erosion performance:

(i) particle characteristics - size, shape, chemistry velocity and impact angle

(ii) oxide scale properties - composition and mechanical properties (especially ductility)

(iii) component material properties - mechanical behaviour, particularly ductility

(iv) temperature and environment

Restall and Stephenson (35) have reported the erosion behaviour of aluminized superalloys and have shown that the impact damage monphologies can be considered in terms of three temperature regimes relating to the oxide scale plasticity and the ductile - brittle transition temperature (DBTT) of the aluminide coating:-

(i) at <800°C, when the coating is below the DBTT and the oxide scale plasticity is low, the erosion exhibits a brittle response whereby oxide is mostly removed down to the oxide scale-coating interface.
(ii) at 800°-900°C, where the coating is above the DBTT but oxide plasticity is only slightly improved by temperature, the impact damage monphology is dependent upon scale thickness. Thick

scales protect the coating, but due to limited plasticity, fracture and oxide removal is observed. In contrast, thinner more plastic scales can accommodate a small degree of deformation before fracture and therefore are somewhat more protective.

(iii) at > 900°C, where the coating is above the DBTT and the oxide plasticity is markedly increased, similar behaviour is observed.

The removal or damage of protective oxide scales evidently has marked implications with reference to oxidation and corrosion processes. In the case of oxidation, the rate of attack will be markedly increased, causing denudation of oxide-forming elements from the material. Similarly, in the case of processes such as hot corrosion, the incubation period before the onset of attack will be decreased and the rate of degradation enhanced.

Thus erosion and oxidation/corrosion intractious will increase the rate of materials degradation, hence decreasing component lifetimes.

4.   **SURFACE COATING TECHNOLOGY**

It is now generally accepted practice to apply surface coatings to high temperature components in gas turbine (36, 37). The principal reasons for coating utilization are (37):

(i)   to ensure that the component is capable of operating for the design lifetime (i.e. coatings to combat oxidation, corrosion and erosion).

(ii) to permit increased component operating temperatures (i.e. thermal barrier coatings)

(iii) to maintain component shape

4.1   CLASSIFICATION OF COATINGS FOR SURFACE
       PROTECTION IN GAS TURBINE

The coating systems currently in use in gas turbines for the protection of vanes and blades against environmental degradation processes can be divided into three groups:-

1.   DIFFUSION COATINGS

These include aluminising, chromizing and silicising. The chromized and siliconized coatings are limited to temperatures below 950°C since both systems form volatile oxide species at higher temperatures, which afford no protection against further oxidation or corrosion, but can however, be successfully used at lower temperatures. Since turbine vanes and blades operate at material temperatures to 1100°C, aluminising is by far the most widely used of these coating systems.

Aluminized coatings are applied to components using a variety of techniques based upon chemical vapour deposition (CVD). The techniques include pack cementation, slurry cementation and vapour phase aluminising (38).

The aluminide coatings are formed by interdiffusion between the depositing Al and the substrate alloy, resulting in the formation of an intermetallic coating comprising NiAl (or CoAl on Co-base alloys). Two principal process routes are available, termed 'low' and 'high' activity. The low activity process is characterized by the outward diffusion of Ni (or Co) from the substrate to form NiAl and occurs at temperatures of ca. 1050°C, whilst the high activity process is a result of the inward diffusion of Al and lower temperatures, forming Al-rich NiAl and Ni$_2$Al$_3$ which is brittle and must be transformed to NiAl by a post-coating heat treatment.

The structure of aluminized coatings has been described in detail by Slattery (39).

Figure 5 shows a high activity pack aluminized coating on IN100 (Ni-base superalloy) in the coated/heat treated condition.

A development of aluminide coatings in order to achieve increased degradation resistance has resulted in the so-called 'modified aluminide

13

systems (40) which may be used when an improvement over aluminising is required. Typical additions include Pt, Cr, Si, Ta etc. The processing of the modified aluminides may involve a variety of techniques, including electroplating or PVD for the initial metallic layer (Pt, Cr etc.) and subsequent pack or vapour aluminising (high or low activity), followed by complex heat treatment cycles in order to produce the correct microstructure and composition, both of which markedly influence the performance of the modified aluminide. Pt-aluminides are by far the most widely used of the modified aluminides and careful selection of their microstructure and composition results in a significant improvement over aluminides in a gas turbine operating regime. The development of Pt-aluminides has been described in detail by Bettridge and Wing (41) and has been the subject of several studies by Boone and Streiff (42).

OVERLAY COATINGS

Overlay coatings are based upon the M-Cr-Al-X system where M is Ni, Co, Ni-Co or Co-Ni and X is an active element (or elements) such as Y, S, Si, Ta, Hf etc (38, 43).

In binary Ni-Al and Co-Al alloys, large levels of Al are required to form and maintain a protective Al_2O_3 scale. However, the addition of 10% Cr drops the required level to around 5% Al. The composition of the M-Cr-Al system can therefore be selected to produce optimum degradation resistance although variation of the Al content also influences the ductility of the coating system (44). The active element additions enhance the adhesion of the protective oxide scale on the coating and decrease the degradation kinetics (38, 43, 45). Active elements such as Y, Hf, Ta, Si etc, whilst causing similar overall effects on the oxidation of M-Cr-Al systems, function via different mechanisms with the magnitude of their influence sometimes being temperature dependent (45). Since turbine vanes and blades encounter a range of temperatures it can be most beneficial to include multiple active element additions in order to take advantage of their synergistic effects and so maximize the degradation resistance of the coating.

Overlay coating systems comprise B-NiAl (or B-CoAl) and a Ni (or Co) matrix containing elements such as Cr in solution. During post-coating heat-treatment and service, coating-substrate interdiffusion occurs with alloying elements such as Ni, Co, Cr, Ti, W, Hf, C etc. diffusing into the coating. The coating-substrate interdiffusion zone usually contains carbides, 'fingers' of NiAl (or CoAl) and in some cases discrete particles of sulphides and oxides formed by the reaction of active elements such as Y with alloy S or O impurities.
A typical overlay, CoNiCrAlY, is shown in Figure 6 on the Ni-base alloy IN100 in the coated/heat treated condition.

The principal deposition processes for overlay coatings are plasma spraying techniques (low pressure plasma spraying, argon-shrouded plasma spraying) and PVD processes (38, 46). Other techniques include composite plating (47) and laser melting (48).

THERMAL BARRIER COATINGS

Major increases in the power and efficiency of gas turbine engines can be achieved by raising the turbine operating temperature (49). Improvements in superalloy technology can achieve only relatively small increase in operating temperatures for components such as nozzle guide vanes and blades, whilst the use of enhanced active component cooling (to maintain a 'reasonable' component temperature) may utilize so much compressor bleed air that there would be a significant shortfall in the projected increase in engine efficiency (50).

The TBC concept provides a means of raising the operating temperature of components whilst using currently available superalloy component materials. The thermal barrier coating should satisfy several requirements, including high melting point, low thermal conductivity, a flame reflectance higher

15

than the substrate material, thermal expansion similar to the substrate, ability to withstand all strain temperature conditions encountered by the component, resistance to oxidation, corrosion and erosion and should maintain the required component aerodynamics.

In a gas turbine, TBC systems can be used to (50):

(i) provide increased turbine gas inlet temperatures whilst maintaining a constant component temperature, thus increasing engine efficiency and performance;

(ii) decrease the use of cooling air in components at constant gas and component temperatures, thus increasing engine efficiency;

(iii) decrease alloy/component temperature at constant gas operating temperatures, thus increasing component life and engine reliability.

The design philosophies for TBC systems applied to vanes and blades have been described by Bennett et al (51) and Miller (52). A safe philosophy involves TBC utilization on a component designed to give an acceptable life without a TBC, thereby achieving a prolonged life under 'normal' operating conditions. A more efficient philosophy is to assume that the component will suffer restricted life in the absence of the coating. This approach permits the use of increased turbine inlet temperatures and/or decreased utilization of component cooling air, thereby improving engine performance.

The most commonly used TBC system is a two layered coating comprising a low thermal conductivity ceramic deposited over a M-Cr-Al-X (overlay system) bond coat. The bond coat provides oxidation and corrosion resistance for the substrate, some accommodation of the thermal expansion mismatch between the alloy and the ceramic, and a bond between the alloy and the ceramic.

The most widely used ceramics are based upon ZrO_2, which exhibits thermal expansion characteristics close to superalloys. In order to avoid deleterious phase changes in the ZrO_2 during thermal cycling exposure, particularly the tetragonal-monoclinic transition, which is characterised by a significant volume change and consequently, cracking of the ceramic, it is usual to add stabilizing species such as Y_2O_3, CeO_2 and MgO (38, 50). The chemistry of TBC systems has been described in detail in the literature (53, 54). The most widely used system is based on $ZrO_2-Y_2O_3$, with the partially stabilized $ZrO_2-8\%Y_2O_3$ providing the most promising ceramic layer.

The ceramic layer is generally deposited by plasma spraying (55, 56) or PVD techniques (57, 58). Deposition processes for M-Cr-Al-X systems have already been outlined.

Figure 7 shows a TBC system comprising an argon shrouded plasma sprayed CoNiCrAlY bond coat and an air plasma sprayed $ZrO_2-8\%$ Y_2O_3 ceramic layer.

4.2   PROPERTY REQUIREMENTS FOR TURBINE VANE AND BLADE COATING SYSTEMS

The properties required by a protective coating for gas turbine vane and blade applications are shown in Table 1 (38, 59) from which it is evident that a wide range of characteristics must be satisfied. The properties must be maintained over all operating conditions encountered by the coated component. The optimization of all properties is clearly unachievable and a more realistic approach involves the development of coatings offering a compromise between the required characteristics.

The characteristics required by a TBC system are listed in Table 2 (50, 60). As in the case of the diffusion and overlay coating systems, optimization of each of the properties for a specific TBC is difficult and selected TBC systems are likely to offer a compromise between the necessary properties.

17

5. DEGRADATION OF HIGH TEMPERATURE COATINGS

5.1 DIFFUSION AND OVERLAY COATINGS

The degradation of aluminide, Pt-aluminide and overlay coatings occurs via (38):

(i) removal of coating constituents by oxidation and corrosion processes, which may under certain conditions be enhanced by erosion

(ii) substrate – coating interdiffusional processes, generally resulting in the diffusion of alloy elements into the coating and aluminium from the coating into the alloy, thereby influencing the performance of the coating.

(iii) strain induced damage processes arising from the thermomechanical strains encountered by the coated component during service.

Figures 8 and 9 show aluminide, double aluminide and overlay systems after an in-service engine test. Whilst the coatings have all afforded satisfactory protection, the MCrAlX overlays have clearly provided the most promising results.
Figure 10 shows the importance of evaluating coating systems on the actual alloy to be used in operation. The figure clearly shows the different behaviour of a Co-Cr-Al-Y overlay coating on two blade alloys tested under the same conditions, thereby highlighting the fact that coating behaviour is affected by the selection of substrate materials.

5.2 THERMAL BARRIER COATINGS

DEGRADATION BY GASEOUS AND DEPOSITED CONTAMINANTS

The ZrO_2 – Y_2O_3 ceramic layer of a TBC is a fully oxidized system, so oxidation attack is limited to the underlying M-Cr-Al-X bond coat system. Under such conditions the function of the ceramic is to decrease the operating temperature of the underlying M-Cr-Al-X / superalloy system, thereby decreasing on oxidation rate of the bond coat. However it has been reported (61) that the growth of only some 3-5 um of oxide on the bond coat surface (i.e. the M-Cr-Al-X/PYSZ interface is sufficient to cause spallation of the ceramic.

In SO_x - containing environments, although ZrO_2 is resistant to sulphidation, the stabilizing phases such as Y_2O_3 (or CeO_2, MgO etc) can all be sulphated, resulting in destabilization of the ZrO_2 and consequently, phase transistions with associated volume changes, consequent ceramic cracking and failure (62, 63). Deposited contaminants such as Na_2SO_4, have also been shown to cause TBC failure, with permeation of the molten salt through the ceramic resulting in bond coat attack, the formation of degradation products and consequent ceramic spallation (64). It has also been reported (65) that the Y_2O_3 stabilizing phase in $ZrO_2-Y_2O_3$ is susceptible to Na_2SO_4 attack according to the reaction:

$$Y_2O_3 \quad + 3SO_3 \quad \rightleftharpoons \quad 2Y^{3+} +$$
$$\text{(TBC)} \quad \text{(Na_2SO_4)} \quad \text{(Na_2SO_4)}$$

$$3SO_4{}^{2-}$$
$$\text{(Na_2SO_4)}$$

---- 9

In addition to weakeninq the M-Cr-Al-X / ceramic adhesion and causing interfacial delamination, the growth of oxidation or corrosion products at the interface induces strains in the overlying ceramic, thus enhancing ongoing thermomechanical strain - induced failure mechanisms, leading to delamination within the ceramic itself (52, 60).

Figure 11 shows a TBC system on a Ni-base superalloy after engine testing, showing degradation of the bond coat and spallation of the ceramic.

DEGRADATION BY EROSION PROCESSES

A study was performed at the University of Cincinnati under Professor W Tabakoff and F C Toriz (Rolls-Royce) to investigate the erosion behaviour of a range of TBC systems including both plasma sprayed and PVD ceramic layers. The high temperature erosion rig has been described in detail by Tabakoff and Wakeman elsewhere (66). The test variables used in the study are listed in Table 3. Figures 12-14 show the influence of particle impingement angle, A; particle velocity, V; and temperature, T, upon the erosion behaviour of a range of TBC systems.

The erosion results may be summarized as follows (50, 67):-

(i)   the erosion rate of the TBC systems increased with rising particle impingement angle, increasing particle size and rising temperature (in the tested range).

(ii)   surface finish of the TBC deteriorated with time (particle weight)

(iii) plasma sprayed coatings of the same nominal composition produced by different suppliers exhibited different erosion rates

(iv) PVD ceramics exhibited far superior erosion resistance than equivalent plasma sprayed systems.

6.   CONSEQUENCES OF COATING DEGRADATION

6.1   OVERLAY AND DIFFUSION COATINGS

The effects of environmental degradation processes such as oxidation, corrosion and erosion include:

(i)   coating/alloy material wastage i.e. loss of component shape and decreased life

Two turbine blades removed from engine service after long exposure are shown in Figure 15. The blade surfaces show clear signs of degradation and localised coating removal. Blades in such condition would normally be replaced when detected during periodic engine inspection.

The useful lifetime of coatings depends upon their ability to resist degradation processes such as

oxidation, corrosion and erosion whilst maintaining
mechanical integrity. Metallographic cross sections
of a range of coating systems on engine run turbine
blades are shown in Figure 16, from which it is
evident that although the coatings are being
progressively degraded, the underlying blade
material itself is protected.

Figure 17 shows the effects of erosion by carbon
upon the integrity of a coated turbine blade. The
coating has been totally removed by erosion, thereby
leaving the substrate material unprotected against
oxidation and corrosion processes.

(ii) corrosion processes decrease the fatigue and
creep performance of component materials, thereby
decreasing component lifetimes.

(iii) enhanced degradation may cause rapid
penetration of the corrosion 'front' through an
aerofoil wall section, perhaps reaching the internal
air cooling holes, thereby disrupting the blade
cooling pattern and, consequently, causing the
component to function at a higher temperature,
thereby decreasing the working lifetime.

The penetration of cooling holes in a turbine
aerofoil section by on corrosion front is shown in
Figure 18.

During engine operation the coated superalloys
also suffer mechanical degradation (fatigue, creep,
thermomechomical loading etc) and coatings must
therefore maintain adequate mechanical integrity
during operation. If coating systems crack as shown
in Figure 19, there is a danger of the crack
propagating to the coating/substrate interface where
it may:

(i)   arrest
(ii)   propagate along the interface
(iii) propagate into the substrate component
material thereby decreasing the life.

Whilst the mechanical behaviour of coated component

materials is of major importance it is beyond the scope of this chapter. Readers are, however, recommended to gain an appreciation of this subject by referring to references 68-70.

6.2   THERMAL BARRIER COATINGS

An understanding of the surface finish deterioration of TBC systems is a major consideration for hot section components.

The aerodynamic efficiency of the gas flow over vanes and blades is markedly influenced by their surface finish (71). A plot of aerodynamic efficiency, expressed as a primary loss coefficient, against Reynolds Number for different aerofoil and TBC surface finishes is shown in Figure 20 (72). The deleterious effects of increasing surface roughness on aerodynamic efficiency and the desirability of a smooth surface are clearly evident.

As the TBC surface roughens, the adjacent gas boundary layer becomes more turbulent, causing increased mixing with the mainstream turbine gases, thereby raising the temperature in the immediate vicinity of the blade. This causes an increase in the TBC surface temperature due to enhanced heat transfer from the adjacent gas (50, 73).

In addition, the increased TBC surface roughness also deleteriously affects the thermal performance of the coating. As the surface roughness of the TBC increases, with an associated rise in the coating surface area, the heat transferred into the component from the high temperature gaseous environment is increased (50, 73).

The basic conceptual one-dimensional heat – transfer equation which may be considered on a simplistic basis for a turbine wall-section comprising alloy/TBC is as follows (73):

$$Q_T = \frac{K_w}{Y_W} A_T (T_1 - T_2) \qquad \text{----}10$$

where Q_T is the total quantity of heat transferred per unit time, K_W is the thermal conductivity of the wall, A_T is the total surface area of the wall, Y_W is the wall thickness and T_1 and T_2 are the 'hot' and 'cold' wall temperatures respectively, and are assumed to be constant over their respective surfaces.

Clearly if the surface area A_T is increased and/or the TBC surface temperature, T_1, increases, the total quantity of heat transferred per unit time, Q_T, rises and the component temperature increases. Significant increases in component temperature will result in decreased service lifetimes.

7. CONCLUSIONS

1. Superalloy materials used in hot-section applications in gas turbines are susceptible to oxidation, corrosion and erosion processes which may become life-limiting under certain circumstances.

2. The oxidation, corrosion and erosion processes occurring in the turbine section have been outlined and mechanisms of degradation described.

3. The use of surface coating technology to provide protection for turbine vane and blade materials has been highlighted available coating systems described and their performance discussed with reference to aero engine applications and experience.

4. The consequences of material/coating breakdown in the gas turbine have been described with reference to the effect upon engine component life and performance.

8. ACKNOWLEDGEMENTS

Permission by Rolls-Royce plc to publish this work is gratefully acknowledged. Preparation of the manuscript by Rachel Slade of Rolls-Royce plc is much appreciated.

9. REFERENCES

1.   D L Driver, D W Hall and G W Meetham, 'The Development Of Gas Turbine Materials', Ed. G W Meetham, 1981, Applied Science Publishers, pp 1-30.

2.   C T Sims and W C Hagel (Eds.), 'The Superalloys', 1972, John Wiley and Sons.

3.   F L Versnyder in 'Proc. High Temperature Alloys For Gas Turbines', 1982, D Reidel, pp 1-52.

4.   G W Meetham (Ed.), 'The Development Of Gas Turbine Materials, 1981, Applied Science Publishers.

5.   C T Sims in 'Proc. High Temperature Alloys For Gas Turbines, 1978, Applied Science Publishers, pp 13-68.

6.   J Stringer, 'Hot Corrosion In Gas Turbines', Report 72-08, Metals And Ceramics Information Center, Cleveland, Ohio, USA, 1972.

7.   T N Rhys-Jones, 'Protective Oxide Scales On Superalloys And Coatings Used In Gas Turbine Vane And Blade Applications, Mats. Sci. and Tech., May 1988, Vol. 4 pp 421-430.

8.   J Stringer, Materials Sci. and Technology, 1987 Vol. 3, No 7, pp 482-493.

9.   I G Wright 'Oxidation Of Iron -, Nickel- and Cobalt-base Alloys', Report 72-07, Metals And Ceramics Information Centre, Cleveland, Ohio, USA, 1972.

10.   A R Nicholl, Proc. Course On High Temperature Materials And Coatings, Finland, June 1984, Continuing Education Institute.

11.   F H Stott, G C Wood and M G Hobby, Oxid. Met. 1971, 3, (2), p 103.

12.   G C Wood and F H Stott, Oxid. Met. 1971, 3, (4) p 365.

13. G R Wallwork and A Z Hed, Oxid, Met, 1971, 3, (3), p 213.

14. C S Giggins and F S Pettit, 'Hot Corrosion Degradation Of Metals And Alloys:- A Unified Theory', Report F R 11545, Pratt and Whitney Aircraft Corporation, East Hartford, CT, USA, 1979.

15. J A Goebel, F J Felton and F S Pettit, 'Proc. Conf. Gas Turbine Materials In A Marine Environment,' Castine, Maine, USA, July 1974, Metals And Ceramics Information Centre, Cleveland, Ohio, USA.

16. R A Rapp and K S Goto in 'Proc. Symposium On Fused Salts' (Ed. G Braunstein and R Selman) 1979, The Electrochemical Society, USA.

17. R F Reising and D P Krause, Corrosion, 1974, 30, pp 513-521.

18. A J B Cutler, J App. Electrochem, 1971, 1, pp 19-26.

19. R L Jones and S T Gadomski, J Electrochem. Soc., 1971, 124, pp 1641.

20. R L Jones, K H Stearns and S T Gadomski in 'Proc. 3rd UK/US Conf. Gas Turbine Materials In A Marine Environment', Bath, Sept. 1976, MOD(UK).

21. R L Jones, K H Stearns, M L Deanhardt and J C Halle in 'Proc. 4th UK/US Conf. Gas Turbine Materials In A Marine Environment', Annapolis, MD, USA, June 1979, US Navy.

22. J A Goebel and F S Pettit, Metall. Trans., 1970, 1943, p 3421.

23. D W McKee and G Romeo in 'Proc. 2nd UK/US Conf. Gas Turbine Materials In A Marine Environment', MCIC Report 75-27, Castine, Maine, USA, July 1974.

24. P Hancock, R C Hurst, J Johnson and M Davies in 'Deposition And Corrosion In Gas Turbines', Eds. A B Hart and A J B Cutler, 1973, Applied Science Publishers, pp 153-157.

25. J F Conde and C G McCreath, 'Proc. Conf. On Behaviour Of High Temperature Materials In Aggressive Environments', 1980, The Metals Society, London, UK, pp 497-512.

26. P Hancock and J R Nicholls, High Temp. Techn., Aug 1982, 1, (1) pp 3-17.

27. C J Spengler, S Y Lee and S T Schreirer, 'Proc. Conf. Advanced Materials For Alternative Fuel Capable Directly Fired Heat Engines,' Castine, Maine, July/Aug 1979, EPRI, USA.

28. N S Bornstein and M A DeCrescente '2nd UK/US Conf. Gas Turbine Materials In A Marine Environment', Castine, Maine, USA, July 1974, Metals And Ceramics Information Centre, Cleveland, Ohio, USA.

29. T N Rhys-Jones, J R Nicholls and P Hancock, Corrosion Science, 1983, 2, (2), pp 139-149.

30. W J Hefner and F D Lordi, Report No. GER-3119, 1979, General Electric Co., USA.

31. T N Rhys-Jones, J R Nicholls and P Hancock, in 'Prediction Of Materials Performance In Plants Operating With Corrosive Environments', 1987, Ellis Horwood, Eds. J R Nicholls and J E Strutt, pp 289-311.

32. J E Restall in 'The Development Of Gas Turbine Materials', Ed. G W Meetham, 1981, Applied Science Publishers, pp 259-293.

33. J C Galsworthy, J E Restall and G C Booth in 'Proc. Conf. High Temperature Materials For Gas Turbines', Ed. R Brutenaud et al, 1982, D Reidel, pp 207-235.

34. P Hancock, A R I Marie and J R Nicholls, Proc. Conf. Advanced Materials For Alternative Fuel Capable Directly Fired Heat Engines', Castine, Maine, July/Aug, 1979, EPRI, USA.

35.   J E Restall and D J Stephenson, 'Proc. Int. Symposium On High Temperature Corrosion', Marseilles, France, July 1986, European Federation Of Corrosion.

36. T N Rhys-Jones, 12th International Thermal Spraying Conference, 4th-9th June 1989, London, UK, Paper 90, The Welding Institute, Cambridge, UK, 1989.

37.   L Peichl, J Wortmann and H J Ratzer-Scheibe, in 'High Temperature Alloys: Their Exploitable Potential', Eds. J B Marriott, M Merz, J Nihoul and J Ward, Applied Science Publishers, Feb. 1988.

38. T N Rhys-Jones and D F Bettridge in 'Proc. Conf. High Temperature Materials And Processing Techniques For Structural Applications', Paris, France, Sept. 1987, ASM Europe, pp 129-158.

39.   G F Slattery, Metals Technology, 1983, Vol 10 pp 41-51.

40. G Lehnert and H W Meinhardt, Dew. Tech. Ber., 11, (4), pp 236, 1971.

41.   D F Bettridge and R G Wing, 31st ASME International Gas Turbine Conference And Exhibit, June 1986, Dusseldorf, F.R.C.

42.   R Streiff and D H Boone in 'Proc. Symposium On The Reactivity Of Solids, Aug 1984, Electrochemical Society USA.

43.   T N Rhys-Jones, 'Surface Coating Technology For Critical Gas Path Components In Aero Engines', 2nd Parsons International Turbine Conf: Materials Development In Turbomachinery Design, Sept. 1988, Churchill College, Cambridge, UK., The Institute of Metals, 1989, in press.

44. G W Meetham, Mats. Sci. and Techn., 1986 Vol 2, No. 3, pp 290-294.

45.   T N Rhys-Jones, Corrosion Science, 1989, in press.

46. J R Nicholls and P Hancock, Industrial Corrosion, July 1987, pp 8-18.

47. E Kedward, C A Addison and F Honey, UK Patent 2-014-189 (1979) and US Patent 4-305-792 (1981).

48. J E Restall and M I Wood, Mats. Sci. and Techn. 1986, Vol 2, 3, pp 225-231.

49. H C Cohen, G F C Rogers and H I Saravanamuttoo, 'Gas Turbine Theory', (2nd Edn.), 1972, Longman, Chapter 1.

50. T N Rhys-Jones and F C Toriz, High Temp. Techn. Vol 7, No 2, May 1989, pp 73-81.

51. A Bennett, F C Toriz and A B Thakker, Surface Coatings And Technology, 1987, 32, pp 359-375.

52. R A Miller, Trans. ASME, Oct, 1987, 109, pp 448-451.

53. R A Miller, J L Smialek and R G Carlick, 'Advances In Ceranmics Vol 3: Science And Technology Of Zirconia', Eds. A H Heuer and L W Hobbs, 1981, American Ceramics Society, pp 241-253.

54. T E Mantkowski, D V Rigney, M J Froning and N Jayaraman', Proc. Advanced Materials Research and Development For Transport, Strasbourg, 26th-28th Nov. 1985, European Materials Research Society.

55. R Burgel and I Kvernes in 'Proc. Conf. High Temperature Alloys For Gas Turbines And Other Applications', Liege, Belgium, 6th-9th Oct., 1986, D Reidel, 1986 pp 327-356.

56. R Bauer, R Burgel and K Schneider, COST 501-D/28, Annual Report, Brown Boveri and Cie AG, Manheim, FRG, 1983.

57. R F Bunshah, 'Materials Coating Techniques', AGARD, LS-106, 1980, Chapter 1.

58. A S James, K S Fancey and A Matthews, Surface Coatings And Technology 1987, 32, pp 377-387.

59.   P Hancock and J R Nicholls, 'Advanced Workshop On Coatings For Heat Engines', April 1983, Maratea, Italy.

60.   T N Rhys-Jones 'Thermal Barrier Coatings For Aero Engines', Proc. Advances In Coatings And Surface Treatments For Corrosion And Wear Resistance, Sept. 1988, Newcastle-upon-Tyne, UK, in press.

61.   R V Hillery, B H Pilsner, R L McKnight, T S Cook and M S Hartle, 'Thermal Barrier Coating Life Prediction Model Development-Final Report', Nov. 1988, NASA Report CR 180807.

62. A S Nagelberg, J Electrochemical Society, 1985, 132, pp 2502.

63.   P Hancock, 'Degradation Processes For Ceramic Coatings', Proc. Conf. Advanced Materials Research And Development For Transport, 26th-28th Nov., 1985, pp 163-179.

64.   D W McKee and P A Siemers, Thin Solid Films 73, 1980, pp 439-445.

65.   R H Barkalow and F S Pettit, 'Proc. 1st Conference On Advanced Materials For Alternative Fuel Capable Directly-Fired Heat Engines, July/Aug, 1979, Castine, Maine, USA, US Dept Of Energy 1979, pp 258-269.

66.   W Tabakoff and T Wakeman, ASTM-STP664, 1979 pp 123-125.

67.   F C Toriz, 'Erosion Behaviour Of Thermal Barrier Coatings', 40th Pacific Coast Meeting of the American Ceramic Society, 1-4 Nov. 1987, San Diego, USA.

68.   T N Rhys-Jones and T P Cunningham, 'The Influence Of Surface Coatings On The Fatigue Behaviour Of Aero Engine Materials', Surface Coatings and Technology (submitted for publication).

69.   M I Wood, 'The Mechanical Properties of

Coatings And Coated Systems', 2nd Int. Symposium On High Temperature Corrosion Of Advanced Materials And Coatings, 1989 (May), Les Embiez, France, in press.

70.   A Strang and E Lang, 'High Temperature Alloys For Gas Turbines', Liege, Belgium, Oct. 1982, D Reidel Pub. Co. 1982 pp 469.

71.   K Bammert and H Sandstede, ASME Paper 80-GT-80, March 1980.

72.   R M Watt, J L Allen, N C Baines, J P Simons and M George, ASME Paper 87-GT-233, 1987.

73.   D   Butterworth, 'An Introduction To Heat Transfer', British Standards Institution And Council of Engineering Institutions, UK, Oxford University Press, 1977.

TABLE 1

PROPERTY REQUIREMENTS OF COATING SYSTEMS FOR GAS TURBINE AEROFOIL APPLICATIONS

PROPERTY	REQUIREMENTS
Corrosion and oxidation resistance	o initial rapid formation of a thin, uniform, adherent and continuous protective oxide film o slow subsequent rate of scale growth o highly stable and adherent scale o high concn of scale-forming elements in coating o acceptable rate of corrosion/oxidation
Erosion resistance	o ductile and adherent oxide scale o moderate coating ductility
Coating and alloy interfacial stability	o low rates of diffusion across interface o minimum compositional changes, particularly w.r.t brittle phases
Coating adhesion	o matched/similar coating and substrate properties o clean alloy/coating interface
Mechanical properties	o ability to withstand all strain-temperature cycles encountered by the component during service o appropriate coating ductility o little or no effect upon substrate properties
Aerodynamic properties	o best possible surface finish o acceptable thickness and uniformity on aerofoil o minimum loss of surface smoothness during service
Coating process	o optimised for composition, structure, thickness (and thickness distribution if required) o ability to coat complex shapes eg aerofoils o cost effectiveness

TABLE 2

PROPERTY REQUIREMENTS FOR A THERMAL
BARRIER COATING IN AERO ENGINES

Low thermal conductivity

Thermal expansion close to superalloys

Thermal shock resistance

Thermal stability

Thermal fatigue resistance

Thermomechanical fatigue resistance

Corrosion resistance

Erosion resistance

Oxidation resistance

Low stiffness

Good fracture toughness

Low weight

Good adhesion to substrate

Good surface finish

Ability to coat complex shapes (e.g. aerofoils)

Good thickness and coating distribution control

TABLE 3

TEST VARIABLES USED IN EROSION STUDY

Variable	Symbol	Tested Range
Temperature	T	$260-815^{\circ}C$
Particle Velocity	V	$122-305\ ms^{-1}$
Impingement Angle	A	$20^{\circ} - 90^{\circ}$
Particle Size	S	$8-130\ um\ Al_2O_3$
Surface Finish	R	$10-80\ um$

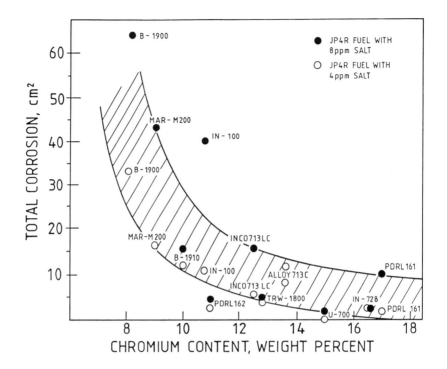

Figure 1   The influence of alloy Cr content on the corrosion behaviour of a range of superalloys in the temperature range 700°-1100°C (Ref 6).

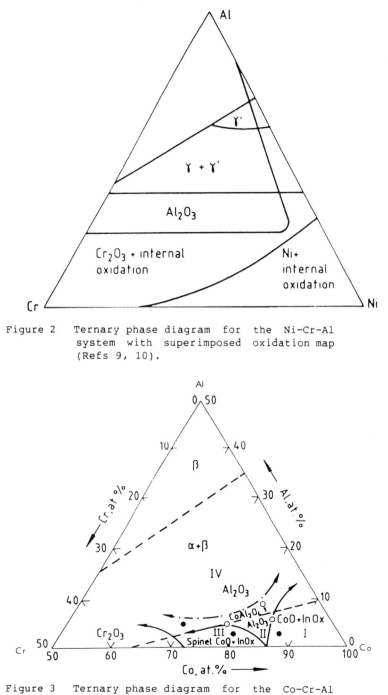

Figure 2   Ternary phase diagram for the Ni-Cr-Al
system with superimposed oxidation map
(Refs 9, 10).

Figure 3   Ternary phase diagram for the Co-Cr-Al
system with superimposed oxidation map
(Ref 13).

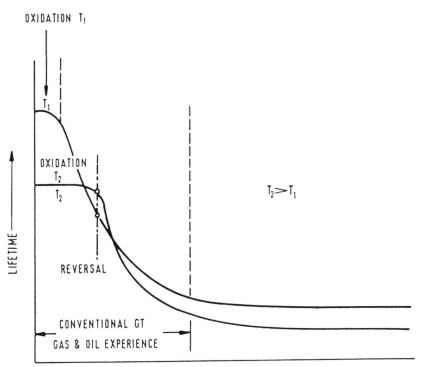

OXIDATION T_1

T_1

OXIDATION T_2

T_2

$T_2 > T_1$

LIFETIME

REVERSAL

CONVENTIONAL GT
GAS & OIL EXPERIENCE

Na_1K CONCENTRATION IN COMBUSTION PRODUCTS

Figure 4   Effects of fuel Na and K content on the
            life of hot section components in a gas
            turbine (Ref 30).

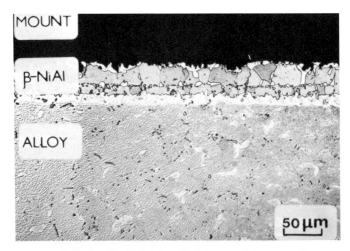

MOUNT

β-NiAl

ALLOY

50 μm

Figure 5   High activity aluminide coating system on
            IN100 (Ni-base superalloy).

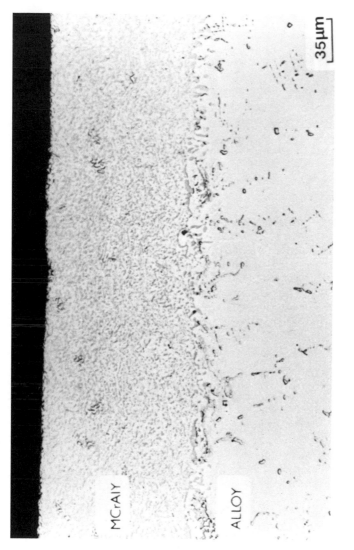

Figure 6   Argon   Shrouded   Plasma   Sprayed
Co-Ni-Cr-Al-Y Overlay Coating system on
IN100 (Courtesy of Union Carbide (UK)
Ltd.).

Figure 7   Thermal Barrier Coating System Comprising
argon shrouded plasma sprayed M-Cr-Al-Y
bond coat and air plasma sprayed
ZrO_2-8% Y_2O_3 ceramic top coat
(Courtesy of Union Carbide (UK) Ltd).

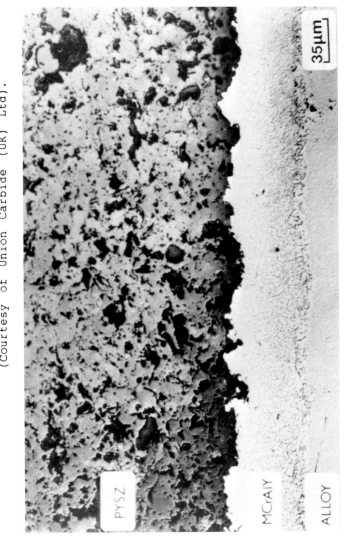

PYSZ

MCrAlY

ALLOY

35μm

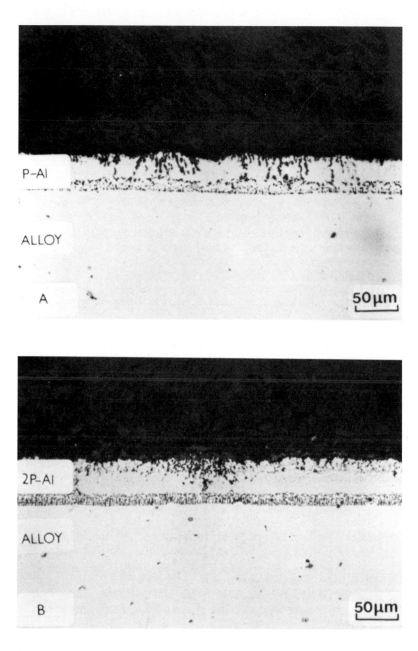

Figure 8   (A) Aluminised and (B) double aluminised
turbine   blade   sections   after   engine
testing.

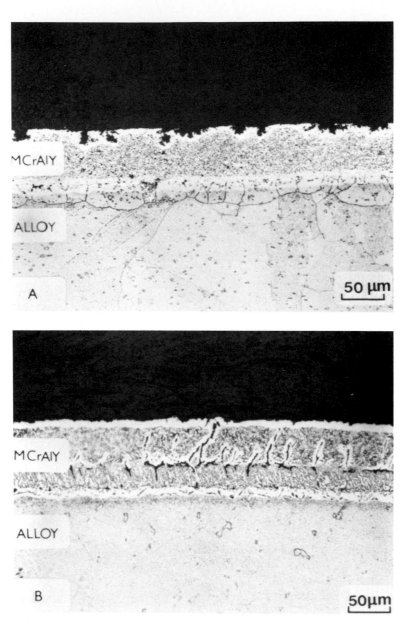

Figure 9   Overlay coated turbine blade sections
            after engine testing (same test as
            aluminide and double aluminide coated
            blade sections in Figure 8).

40

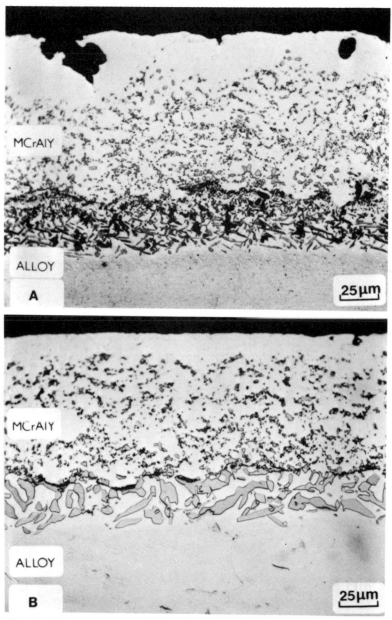

Figure 10   Oxidation   behaviour   of a   Co-Cr-Al-Y
overlay   coating   on   two   different
superalloys (A and B), highlighting the
importance   of   coating-   substrate
interactions   on   the   degradation
resistance of surface coatings.

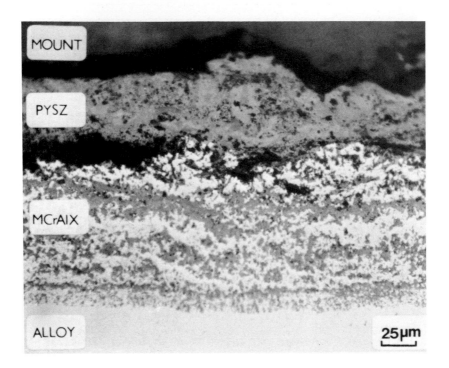

Figure 11   Thermal Barrier Coating system on a
Ni-base superalloy substrate after
prolonged engine testing showing bond
coat degradation and spallation of the
ceramic.

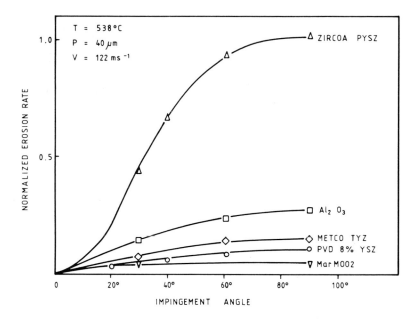

Figure 12 Influence of particle impingement angle on the erosion behaviour of TBC systems, Al_2O_3 and MarM002 ($T=538^{\circ}C$, $V=122ms^{-1}$, $P=40$ um).

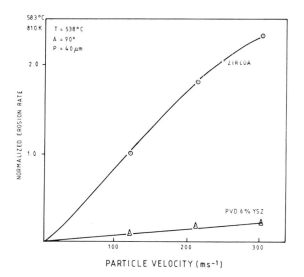

Figure 13 Influence of particle velocity on the erosion behaviour of plasma sprayed and PVD thermal barrier coatings ($T=538^{\circ}C$, $P=40$ um, $A=90^{\circ}$).

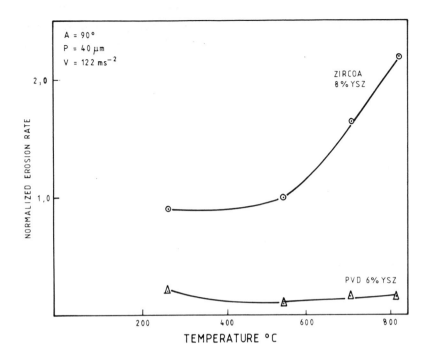

Figure 14   Influence of temperature on the erosion
            behaviour of thermal barrier coatings
            (P=40um, A=90°, V=122ms⁻¹).

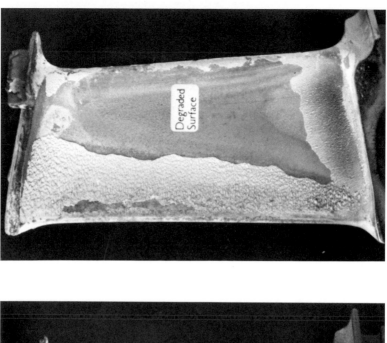

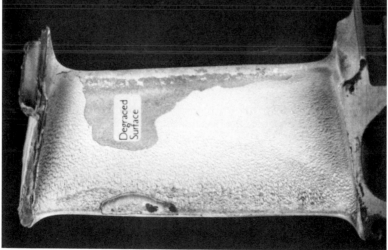

Figure 15   Turbine blades removed from an engine
after   prolonged   exposure,   showing
extensive   degradation   on . aerofoil
surfaces.

45

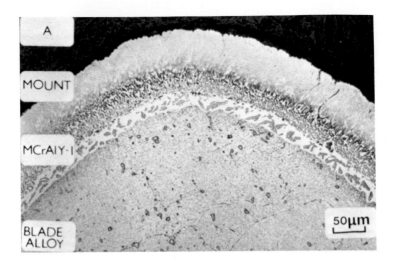

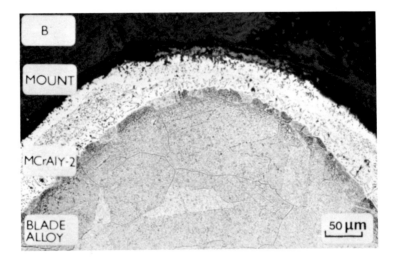

46

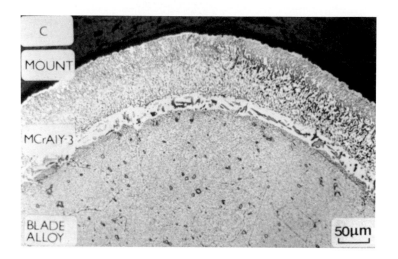

Figure 16   Cross-section views of coated turbine
blades after prolonged operation in an
aero engine (a) MCrAlY-1 (b) MCrAlY-2
(c) MCrAlY-3.

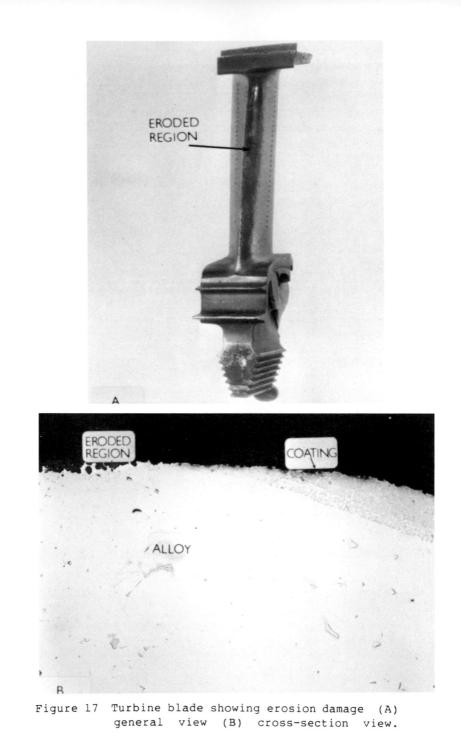

Figure 17   Turbine blade showing erosion damage   (A)
general   view   (B)   cross-section   view.

48

Figure 18 Heavy corrosion damage of a turbine blade
resulting in breaching of air cooling
holes.

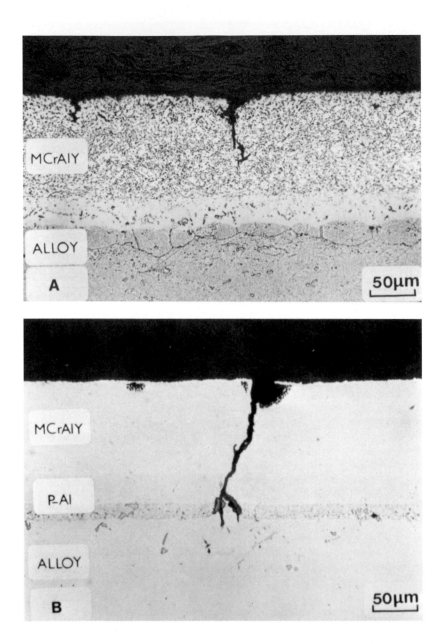

MCrAlY

ALLOY

A   50μm

MCrAlY

P-Al

ALLOY

B   50μm

Figure 19 Mechanical breakdown of a coated turbine
blade showing (A) cracking of coating and
(B) propagation of a crack from the
coating into the alloy substrate.

50

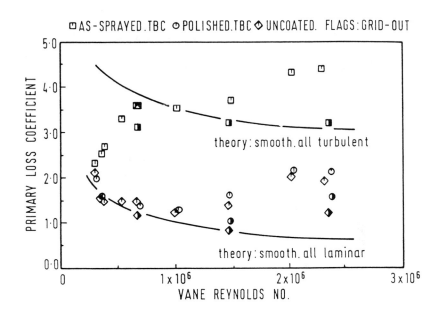

Figure 20   The Influence of Surface Roughness On the 2D aerodynamic efficiency of a TBC-coated turbine blade.

2: Corrosion control in power plants
D B MEADOWCROFT

Barry Meadowcroft is at the Central Electricity
Research Laboratories of the Central Electricity
Generating Board.

1.   INTRODUCTION

The purpose of this paper is to illustrate the
importance to the power generation industry of the high
temperature surface stability of the materials of
construction. Apart from renewables (hydroelectricity,
wind power, etc.), all commercial central electricity
production is by the conversion of high temperature
heat energy to electrical energy via mechanical energy
– a rotating turbine. As all high temperature
materials rely on a surface layer for their long term
stability, the control of high temperature corrosion is
a critical requirement for the power generation
industry.

Steam production is generally the intermediate
stage to drive the turbine, for both fossil and nuclear
fuels, except for gas turbines which convert the heat
energy directly. Thus, in a gas cooled reactor, the
reactor coolant is used to raise steam in "boilers",
just as in a coal or oil fired "boiler" the heat of
combustion of the fuel is used to raise steam. The
steam turbine and electrical generator can thus be
identical for both fossil and nuclear fuels if the
steam is produced to the same pressure and temperature,
as happens with Advanced Gas Cooled Reactors (AGRs).
Schematic diagrams of coal-fired fossil power plant and

nuclear gas cooled reactor power plant are shown in
Fig. 1 and 2 respectively.

It is not possible to discuss all the high
temperature corrosion topics relevant to power
generation in a single paper. A selection has
therefore been made to show the wide range of
conditions which have to be contended with in operating
a reliable and efficient power plant, and the different
corrosion mechanisms involved. It will also be clear
from the paper that materials selection for a
particular plant must always be economic for a
particular application, rather than necessarily the
technical "best".

The paper can be divided into two main parts:
Magnox and AGR gas cooled reactors and their associated
boilers; and coal- and oil-fired boilers. The
oxidation of the high temperature materials in steam
turbines do not generally cause concern, and some
aspects of the behaviour of materials in steam are
dealt with in the discussion on fossil fired boilers.
Aero gas turbines are the subject of a separate paper
by Rhys Jones[1] in this volume, and whilst there are
some specific land based factors, such as increased
impurity levels in the fuel and sometimes in the
combustion air, that paper deals with the main surface
stability concerns.

2.   GAS COOLED REACTORS

In the UK there have been two generations of gas cooled
reactors, both using carbon dioxide to remove the heat
from the reactor – Magnox and AGR (Advanced Gas Cooled
Reactor). The first generation used enriched uranium
fuel in metal cans made of a magnesium, 0.8% aluminium
alloy – Magnox alloy, which gave its name to the design
– chosen because of its low thermal neutron capture
cross section. Maximum coolant gas temperatures
approached 400°C, and pressures were 10-30 bar,
higher in later stations. The second generation of
AGRs have used uranium oxide fuel in steel cans –
20Cr25Ni – with design outlet gas temperatures up to
670°C and pressures generally of 40 bar. In both cases
the neutron moderator in the reactor has been graphite,
and therefore the composition of the carbon dioxide
coolant has had to be carefully controlled to avoid

excessive oxidation of the graphite, in particular by
having the order of 1% carbon monoxide in the gas. In
this section three aspects of the materials
requirements of gas cooled reactors will be discussed:
carbon steels as structural components in Magnox
stations, 9Cr steels as boiler tubes in AGRs, and, more
briefly, 20Cr25Ni steels as the fuel cans in AGRs.

2.1 Carbon steels in Magnox reactors

One of the major problems which has arisen with Magnox
stations has been corrosion of the structural carbon
steels used around the reactor core and in the boiler
tubes. Two detailed reviews which include the
oxidation performance of carbon steels in Magnox
reactors, have been published by Rowlands et al.[2,3]
Whilst carbon steels oxidise protectively in
carbon dioxide at atmospheric pressure, at high
pressures rapid linear rates occur after an incubation
period of protective oxidation. This behaviour is
shown schematically in Figure 3. The "time to
breakaway" can be several years at Magnox operating
conditions, and has been found to depend on a range of
factors including the silicon content of the steel, its
surface finish, and the moisture and carbon monoxide
concentrations in the gas. In addition, a graphite
based paint used on the steels to prevent atmospheric
corrosion during erection, proved to be a very
effective initiator of breakaway oxidation. The latter
illustrates the importance of considering all possible
factors which could possibly affect subsequent high
temperature corrosion.
The basic mechanism of the corrosion process is
that normally mild steel forms a protective layer of
magnetite in carbon dioxide. The oxide is a single
layer on pure iron in the early stages of oxidation,
but usually the presence of even small amounts of
alloying elements causes a double oxide layer to form,
which grows at both the gas and metal interfaces, as
illustrated in Figure 4. The outer layer is magnetite,
whilst the inner layer contains the alloying elements.
Under certain combinations of temperature, CO
content and gas pressure the gas has a carbon activity
greater than unity. This is the case, for example,
with 1%CO in 40 bar gas below 500°C. Thus the gases

are potentially carbon depositing and carburising, as
well as oxidising. During the oxidation process the
carbon dioxide is reduced to carbon monoxide or to
carbon. When carbon is formed at the metal oxide
interface, it is primarily deposited in the oxide, as
the solubility of carbon in iron is less than 0.1% and
the formation of Fe_3C in the metal is not favoured.
This deposition of carbon in the oxide occurs at high
pressures, and up to 25 vol% of carbon can be found in
the oxide. This prevents oxide-oxide grain contact,
and the formation of a continuous oxide layer, so that
ultimately the protective nature of the scale breaks
down. Breakaway in carbon steels often produces
excresences of oxide which rapidly grow through the
protective oxide layer (Fig. 5), though sometimes the
breakaway scale propagates uniformly.

This form of non-protective oxide growth has an
even more deleterious effect than simply causing loss
of metal section. In breakaway all the oxidation
occurs at the metal-oxide interface, such that in a
restrained crevice the increased volume of oxide
compared with metal cannot be accommodated except by
increasing the size of the crevice. The energy
available in the oxide process is sufficient to move
any restraint, and thus a major problem in Magnox
stations has been the fracture of bolts by oxide growth
at the various interfaces being constrained by the
bolts.

Fig. 6 shows an example of a bolt holding together
a large number of washers – which provide a large
number of crevice interfaces – before and after
oxidation in high pressure carbon dioxide. Visible
extension of the bolt is clearly evident. Close to a
reactor, man access is not possible, and the effects of
breakaway oxidation have required the development of
advanced robotic inspection and manipulative devices to
monitor and replace failed bolts and components. A
permanent restriction on the gas outlet temperature
from the reactors has also resulted, and also that the
coolant gas be kept as dry as possible, requiring
reliable measurements of moisture content at the vppm
level.

Many laboratory and plant investigations have been
carried out so that future oxidation rates can be
reliably predicted. For carbon steels in Magnox
conditions breakaway occurs within a few years, and

therefore it is the prediction of post-breakaway rates which are of most importance for plant. Post-breakaway rates have been obtained from over 5000 components removed from stations, and from 7000 laboratory specimens for a range of gas atmospheres and steel composition, especially silicon content. Further, to extend the data base from stations, "monitoring schemes" have been installed in the stations, to enable large numbers of specimens to be exposed to reactor gas in isothermal autoclaves.

Such a large amount of data has required sophisticated statistical techniques to separate the effects of the various variables, so that equations are now available which predict rates over the range 300-550°C. Decreases in temperature, increases in the silicon content of the steels, and decreases in the water content in the gas, were found to be important variables in determining post-breakaway rates.

It has been found that the experimental results can be best fitted to the data if two temperature regimes are identified. At lower temperatures a rate R_1 is most important, which is dependent on the silicon content of the steel ([Si]wt%) as well as a term (T), dependent on the temperature (TEMP (°C)) where:

$$T = (1/TEMP - 1/633.16)/1.987 \qquad \dots 1.$$

At higher temperatures a rate R_2 is most important which is also dependent on the moisture content in the gas through a term W, where

$$W = \{gas\ pressure\ (psi\ a)\}\{moisture\ (vppm)\}/19485$$

These two rates are combined "in parallel" for any particular temperature to give the actual observed rate R:

$$\frac{1}{R} = \frac{1}{R_1} + \frac{1}{R_2} \qquad \dots 2.$$

where

$$R_1 = 0.07\ [Si]^{-0.49} \exp(-43,500T) \qquad \dots 3.$$

$$\text{and}\quad R_2 = 0.48\ [Si]^{-0.185}\ W^{0.90} \exp(-13,000T) \ \dots 4.$$

An example of the fit between experimental and predicted behaviour using the above equations is shown in Fig. 7.

2.2 9Cr Steels in AGR Boilers

AGR heat exchangers are once through boilers, which means that water enters at the bottom of a tube and exits as superheated steam at about 540°C at the top. The upper sections of the tubes are austenitic alloy (18Cr11Ni, AISI 316), and the inlet sections carbon steel. However, it was considered that the central section where the water evaporation occurs should be 9%Cr1Mo steel, because it was, by then, realised that the temperature was too high for carbon steels, and the stress corrosion resistance of austenitic alloys is too poor for them to be used in evaporative conditions. The 9Cr tubes operate up to about 500°C, and in some AGRs they are finned. The behaviour of the 9Cr steels in AGR conditions has been reviewed by Rowlands et al.[2,3].

Investigations of the oxidation behaviour of 9Cr steels showed that they could also suffer from breakaway oxidation at temperatures of 600°C and above. The requirement was to be able to predict the behaviour at temperatures of interest to reactor operation. Estimates were therefore needed of protective oxidation rates, times-to-breakaway, and post breakaway rates. Protective rates can readily be measured, but times-to-breakaway and post breakaway rates needed further consideration.

The times-to-breakaway at these lower temperatures were predicted by mechanistic considerations and by extrapolation of the high temperature data to be many tens of thousands of hour. To obtain breakaway rates at lower temperatures, samples were exposed at a temperature above 600°C until breakaway was established, and were then "transferred" to lower temperatures. The subsequent corrosion rates thus gave breakaway rates at lower temperatures much more quickly than low temperature isothermal exposures could have done.

As indicated above, mechanistic understanding of the processes involved has proved particularly valuable. In these alloys the carbon can be deposited in the metal to form $M_{23}C_6$ carbides, as well as in the oxide, unlike the situation for mild steels. This metal sink for carbon delays greatly the onset of breakaway, which is not possible until the metal is saturated with carbon. As the deposition rate of carbon is proportional to the oxidation rate, this understanding of the behaviour of carbon means that there tends to be a critical weight gain at breakaway – and thus the time-to-breakaway becomes much longer at lower temperatures where the oxidation rate is lower. Further, carbon will build up more rapidly at corners and edges, because of the increased surface to volume ratio in these regions. Thus such features will go into breakaway more quickly, as illustrated in Fig. 8.

Again, the prediction of long term oxidation behaviour has required a range of sophisticated statistical techniques to analyse the data. In this case all three parts of the process – protective oxidation rate, time (weight gain) to breakaway, and post-breakaway rate – have to be accounted for in the model, as they are all of significant length compared with reactor life. Equations have been developed which fit the experimental data for each regime.

For the protective behaviour, the oxidation rate is assumed to be a power law of the form

$$w = at^b \qquad\qquad \dots 5.$$

where w is the weight gain at time t, and a and b are constants. Typical curves are shown in Figure 9. Complex polynomial expressions have been derived for the values of a and b, in terms of temperature and silicon content of the steel.

For the weight gain at breakaway, a factorial experiment was carried out at 640°C using three CO and three moisture levels. The data were analysed statistically[2], to derive probability functions as a function of weight gain at breakaway as shown in Fig. 10.

For post breakaway rates, isothermal and "transferred" data were analysed. A five term equation for post breakaway rate was derived[2], in terms of temperature, water and CO concentrations. The reduction in post breakaway rates between 600°C and 500°C is seen in Fig. 11.

As long term data become available from lower

temperature samples the accuracy of the extrapolated predictions can be tested. The predictions are generally proving accurate, the main parameter needing further refinement being the estimates of time to breakaway.

To estimate the failure probability of a boiler tube in a station, it is necessary to consider how these various terms can be combined. Because there is no evident correlation between the three parts of the oxidation process - protective rate, weight gain at breakaway, and post breakaway rate - a Monte Carlo procedure is used, in which a random sample of each part is taken to generate a oxidation-time history for the station lifetime. This is repeated many times and the fraction of samples which exceed the failure criterion gives the probability of failure. An additional feature which has to be included is that the experimental samples are small compared with a boiler tube. To accommodate this difference a coupon-to-tube factor is incorporated. It is calculated how many coupons make up a tube, say 500, and the behaviour of a tube is assessed by repeating the calculation 500 times (which can be over a range of temperature), to give each random sample for the final probability factor of a tube.

As the above information has been accumulated it has been fed back into modifications to the design of the AGRs. For instance, the hottest 9Cr tubes are not finned in the later stations. All Magnox and AGR stations have oxidation monitoring schemes with coupons and components exposed to reactor gas, which are examined biannually, to give further improved confidence in the oxidation predictions. The work on 9Cr has defined the limits of utilisation of the alloy, such that only at two stations is there any concern about oxidation causing tube failures before end of life, and satisfactory tube lives can be achieved by optimising the operating parameters.

2.3   20Cr25Ni Steels for AGR Fuel Cans

In AGRs the fuel cans, 1 m long by 14.4 mm diameter, are made of 20Cr25NiNb stainless steel, because of the much higher temperatures they have to operate at

compared with Magnox reactors. The cans have to
operate for up to about 40,000 h at temperatures up to
800°C in normal operation and to deal with higher
temperatures in fault conditions. To satisfy these
criteria and yet minimise the absorption of neutrons,
the cans are only 0.38 mm thick.

If the fuel cans fail, radioactive species would
be admitted into the gas circuit. The oxidation
behaviour has therefore been studied in extreme detail
for normal and fault operating conditions[4].

There are three aspects of particular interest in
the oxidation behaviour of this alloy. First its
general oxidation behaviour; second its tendency to
spall; and thirdly, after spalling, the development of
pits and their healing.

Unlike the ferritic steels discussed above, both
for this and other highly alloyed materials, breakaway
does not occur nor does the total gas pressure affect
the oxidation behaviour. The alloy primarily forms a
protective layer of chromia, with frequently a thin
layer ($\sim$50 nm) of amorphous silica between it and the
metal. Internal oxidation also occurs, and both the
chromia and the internal oxidation layers grow
parabolically. The formation of the protective layer
leads to a chromium concentration gradient in the
metal, such that when the oxide spalls at the metal
interface, the chromium concentration is too low at the
surface for a chromia layer to reform immediately. A
spinel layer therefore forms which grows locally much
faster than unspalled areas (see Fig. 12). Its growth
rate is too fast for the chromium depletion layer to
develop further, and thus the chromium level at the
metal-oxide interface increases until a "healing" layer
of chromia can reform at the interface. The chromium
concentration which allows healing is about 14-18%, and
lifetime assessments of the metal loss associated with
spalling assume, conservatively, that a level of 18.5%
must be reached.

The spalling behaviour is very important, both
because of the associated pit development and loss of
metal section, and because the debris is very
radioactive and thus contaminates the gas circuit.

For an austenitic operating at such temperatures,
the mismatch between the thermal expansion coefficients
of the metal and oxide (that for the metal is about
twice that of the oxide) means that oxide spallation

can occur even for very thin oxides. As the metal cools from its operational temperature the oxide goes into compression and can fail by the mechanisms shown in Figure 13. The results reproduced in Figure 14 show how the temperature fall which can be tolerated by the oxide without spalling decreases with increasing oxide thickness. Detailed mechanisms of the theory of spalling have been quantified based on this alloy[4], extending a model initially proposed to explain spalling of steam side oxide in fossil plants[5]. This model is discussed in detail in Section 3.4, it being mentioned here to demonstrate how a corrosion model developed in one environment can be usefully applied in another.

3.   FOSSIL-FIRED BOILERS

As in the gas cooled nuclear reactors the purpose of the boilers is to convert a hot gas into steam. The major difference is that, whilst in the nuclear case the gas composition is very closely defined, in a fossil boiler the composition of the combustion gases is both very variable and also contains a range of minor constituents which can cause rapid corrosion.

Fossil-fired boilers provide good examples of how the selection of corrosion resistant materials must be balanced by the economic requirements of the plant. Whilst in the hot sections of gas turbines economics allow the selection of very expensive alloys and coatings, in fossil fired boilers the economics significantly limit the selection of materials. Thus we talk about the "control of fireside corrosion" rather than its elimination, and the CEGB have published a handbook on the subject entitled "The Control of Fireside Corrosion in Power Station Boilers"[6].

Both because of the different materials of use and the different mechanisms of corrosion involved, consideration of fossil fired boilers must be separated into two distinct regions, as shown in Fig. 15. There is the furnace chamber where the fuel is burnt, and which is surrounded by tubes in which the evaporation process occurs. Above the furnace where combustion is complete, there is a region where banks of tubes hang into the gas space which are used to superheat the

steam up to 540–565°C dependent on the plant (Fig. 16).
Fig. 15 mentions "reheaters". The superheated steam at
up to 565°C and 160 bar is expanded through the first
stage of the turbine down to about 40 bar, and is then
returned to the boiler to be reheated up to 565°C
again. This greatly improves the efficiency of the
turbine cycle. The scale of a modern boiler is
indicated by the drawing of the truck in the figure.
Another statistic is that a single boiler producing the
steam for a 500 MW alternator burns about 200 tonnes of
coal per hour.

An important factor, which must be realised when
considering the appropriate metal temperatures to be
associated with a given steam temperature, is that the
high gas temperatures encountered in fossil-fired
boilers mean that there are high heat fluxes through
the tubes. In the furnace they can be several hundred
kilowatts per sq. metre, and in the superheater region
can be over one hundred. Such high heat fluxes mean
that the surface metal temperatures are significantly
higher than those of the internal steam or water. Thus
while the steam evaporation temperature in modern
boilers is ∿340°C, the furnace wall tubes typically
operate at surface metal temperatures of up to 450°C as
shown in Fig. 17. Similarly, for current outlet steam
temperatures of 568°C, the maximum surface metal
temperatures are up to 650°C, as shown in Fig. 18.

3.1 Furnace Wall Corrosion

Furnace wall tubes are traditionally carbon steel, as
they operate only at temperatures up to 450°C.
However, on occasion, rates up to 1 mm/1000 h have been
recorded on plant. Although tubes are up to 8 mm
thick, with such a corrosion rate they would not last
even for a year, never mind the three years between
major overhauls, and obviously not the expected station
lifetime.

At one time these high rates were thought to be
due primarily to the effects of the chlorine in UK
coals, which contains about 0.25% chlorine on average
but can be as high as 0.7%. As a result of many
laboratory and plant measurements, it is now considered
that the essential precursor to accelerated furnace
wall corrosion is the occurrence of high levels of

carbon monoxide (1%-10%) near to the tubes due to incomplete combustion.

In laboratory tests it has been found that gas mixtures which simulate the combustion gas – viz 10% CO, 10%H_2O, 0.3%H_2S, balance nitrogen – can give rates comparable with those in plant. However, if SO_2 is used as the sulphur species instead of H_2S, then rapid rates do not occur unless there is sufficient iron catalytic surface in the exposure vessel to cause the SO_2 to react to form H_2S which is the thermodynamically preferred species at temperatures of 400-500°C. The laboratory experiments used SO_2 first as this is the species measured predominantly in plant. However, when the thermodynamic calculations are carried out for the gas temperature of ∿1600°C, it is found that SO_2 is then the preferred species. Thus, it is necessary for the furnace wall conditions to be such as to cause the reaction of SO_2 to H_2S. High CO levels near the furnace walls do not appear to be sufficient to cause rapid corrosion, so it is possible that the, as yet unexplained, effect, is concerned with providing the conditions to convert SO_2 to H_2S.

Whether chloride has some contribution is not yet clear. The original evidence was based on experience at a few old stations with low steam pressure conditions which result in maximum metal temperatures of 375°C. In these cases a linear correlation with chlorine content was found. However, the high corrosion rates sometimes found on modern plant with moderate coal chlorine levels, would need chlorine levels >1% to explain them. Thus, the complete explanation is not yet available.

The plant operator frequently wishes to have a materials solution to severe furnace wall corrosion, because it generally occurs in a limited region of the boiler, and combustion modifications could move the affected area rather than eliminate the corrosion. When one investigates what materials are available to replace carbon steel, it is found that low chromium (<12%) steels do not offer any better corrosion resistance. 18Cr10Ni (Type 304) and particularly 25Cr20Ni (Type 310) steels have been found to give rates (assessed as linear) on plant reduced by a factor of three or more compared with carbon steel. Unfortunately it is not possible to use such alloys as monobloc tubes for evaporator tubes, because of their susceptibility to chloride stress corrosion cracking on the waterside. Chloride can readily be present if there are leaks in the condenser after the turbine.

The solution which has been adopted is to produce tubes hot-coextruded with carbon steel on the inside and Type 310 on the outside - see Fig. 19. Whilst a factor of ten more expensive than equivalent carbon steel tubes, when the costs of installation and reduced outage times are taken into account they have proved an economic way of dealing with many cases of severe furnace wall corrosion. However, as most modern plants currently burn coals with similar chlorine levels, its behaviour with high chlorine coals is not yet established.

There are one or two situations where 310 is not sufficient in current plant, and other possibilities are being considered. Different alloys are being evaluated, and the use of in-situ plasma spraying is being strongly progressed, because if it is successful it has been shown that it would be very economic. It would also be of significant value if cheaper ways of producing coextruded tubes were to be developed.

3.2 Superheater Corrosion

Because the impurities which are important in causing superheater corrosion differ between oil and coal, it is better to consider the two fuels separately. In addition, two features of the oxidation processes on the steam-side of superheater tubes will be discussed.

3.2.1 Coal fired superheater corrosion

In the superheater part of the boiler combustion is complete, free oxygen is present (typically a few per cent) and many volatile corrosive species have been released from the coal. It had been thought that the dominant release was that of sodium chloride which then reacted with the sulphur dioxide to produce sulphate which deposited on the tubes. It was also assumed that the sodium chloride reacted with the ash to release potassium salts which particularly reduce the melting point of the deposited sulphates. Whilst the resultant deposition of sulphates is the same, the discovery that

90% of the chloride is released as HCl during the early part of the combustion process, means that it is now considered that it is the HCl which releases both the sodium and potassium salts.

It is also important that the released sulphur from the coal is present as SO_3 as well as SO_2. The SO_3 affects the stability and melting points of the deposits and the formation of molten compounds which are formed by the reaction of the deposits with the metal oxides. The corrosion products also catalyse the oxidation of the SO_2 to SO_3 so aiding the stability of the molten salts, such that they become molten in the region of 600°C. In addition, as SO_3 is more soluble in the molten salt than oxygen, it is the SO_3 which diffuses preferentially through the molten layer, resulting in a low partial pressure of oxygen and high sulphur partial pressure under the molten layer which allows sulphides to form. Further, the oxide is soluble in the molten salt, and the heat flux through the tube causes a continuous dissolution of the oxide and its precipitation non-protectively on the outside of the protective layer. This mechanism predicts that the corrosion rate will be linear as observed. The solubility of the oxide decreases at higher temperatures leading to a decrease in corrosion rate above 700°C.

In practice, for plant with final steam temperatures of 568°C, the mechanical properties of the available steels means that austenitic alloys of the 18Cr10Ni type (316, 321, 347) have been generally used for the hottest tubes where molten salt corrosion is likely to be important. (In the latest CEGB stations Alloy 1250 (15Cr6Mn10Ni) has been used instead as it has improved high temperature strength, but its corrosion resistance is similar to the '300' series steels). Both from plant experience and laboratory tests the corrosion rate of such materials is found to increase rapidly above 600°C, as shown in Fig. 20. The inset to the figure shows the decrease at higher temperatures found in laboratory tests and from test probes in boilers - classically called the "bell-shaped" dependence. The figure also shows that the corrosion rate depends significantly on gas temperature. This is presumably an indicated effect of heat flux, but no suitable quantitative expression involving heat flux has been found which fits the data.

Empirical relationships only have been found, which show also a linear dependence of corrosion rate on chlorine content in the coal.

It is instructive to examine the form of the empirical derived equation for the corrosion rate of austenitics:

$$R = K \left[\frac{Tg-A}{B}\right]^m \left[\frac{Tm-C}{D}\right]^n [Cl-E] \qquad \ldots 6.$$

where   R is the corrosion rate
               Tg is gas temperature
               Tm is metal temperature
               Cl is the wt% content of chlorine in the coal
and   A, B, C, D, E, K, m and n are constants,

from the point of view of how accurately it enables corrosion rates to be predicted.

It is not possible to know accurately the gas and metal temperature at a given point, and ±50°C in Tg (σ_g), and ±10°C in Tm (σ_m), are reasonable estimates of their variability. Coal chlorine analyses are accurate to about ±0.02% (σ_{Cl}), neglecting the variation from delivery to delivery. These errors in Tm, Tg, and Cl, result in an imprecision in the prediction of a corrosion rate (σ_R) by the above equation which is given by the standard partial differential form:

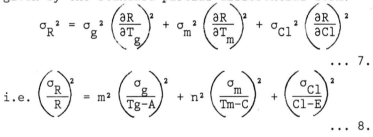

$$\sigma_R{}^2 = \sigma_g{}^2 \left(\frac{\partial R}{\partial T_g}\right)^2 + \sigma_m{}^2 \left(\frac{\partial R}{\partial T_m}\right)^2 + \sigma_{Cl}{}^2 \left(\frac{\partial R}{\partial Cl}\right)^2$$

$$\ldots 7.$$

$$\text{i.e.} \left(\frac{\sigma_R}{R}\right)^2 = m^2 \left(\frac{\sigma_g}{Tg-A}\right)^2 + n^2 \left(\frac{\sigma_m}{Tm-C}\right)^2 + \left(\frac{\sigma_{Cl}}{Cl-E}\right)^2$$

$$\ldots 8.$$

For the estimates in the variability of temperatures and coal chlorine content given above, the actual fit to the data in Figure 13 leads typically to a 25% imprecision in any predicted value of corrosion rate. That is, even if the correlation were 100% accurate (which of course it is not), the best prediction of corrosion rate could be correct only to ±25%. This is because of the strong dependence of corrosion rate on temperature, particularly of the metal, as can be deduced by inspection of Fig. 20.

For superheater tubes, as for furnace wall tubes, the rates found on plant have proved to be excessive for the standard austenitic alloys. As for furnace walls, the steel AISI 310 (25Cr20Ni) has been found to have significantly better corrosion resistance than the standard materials (300 series alloys in this case). However, the standard austenitics were selected for their high temperature mechanical strength and creep resistance, and AISI 310 is not a suitable replacement in this regard. Again, the materials solution has been to use coextruded tube with an outer of AISI 310, but, in this case, the inner is of Alloy 1250 to provide the high temperature creep resistance.

The use of a coextruded tube has a second advantage compared with proving a new monobloc material. For a novel monobloc material, the alloy's creep resistance must first be proved, which takes many thousands of hours of testing, before it can be used in plant. With coextruded tubing no allowance is made in determining the required thickness of the tube layers for any strength contribution from the outer layer. The thickness of the inner layer is chosen to satisfy the mechanical property requirements. Thus coextruded tubing can allow novel corrosion resistant materials to be more quickly tested in plant, as well as allowing otherwise inadequate materials to be utilised.

A further similarity with the furnace wall case is that the benefit factor for AISI 310 is about a factor of three compared with standard alloys, which is not sufficient for all UK plant. An outer layer of 50Cr50Ni alloy has been found to have a benefit factor of at least ten, but it is a two phase material which cannot readily be coextruded in a wrought form. It is possible to produce it using a powder metallurgy route, (Fig. 21), but at much greater cost than the AISI 310/Alloy 1250 combination. This has limited its commercial utilisation greatly, and there is currently an opportunity for a more corrosion resistant alloy than AISI 310 which can be coextruded. Yet again, the additional cost of coextrusion is still a bar to the widest utilisation of such corrosion resistant tubes, and cheaper methods of fabrication would be commercially advantageous. This is a very clear example of a "perfect" solution technically not being acceptable economically.

3.2.2 Oil fired superheater corrosion

The impurities in the heavy oils burnt in power plant
(usually the residual left after refining) are
primarily sodium, vanadium and sulphur. There is also
a lack of the ash. Both factors tend to make
conditions more aggressive. The lack of ash means that
there is no medium to absorb the aggressive salts, such
that concentrations of impurities as low as 5 vppm can
be significant in causing corrosion.

In this case the deposits are sulphate-vanadate
compounds which melt at lower temperatures than those
encountered in coal firing. High vanadium to sodium
ratios are found to cause increased corrosion rates.
The aggressiveness of the conditions in causing
sulphidation means that alloys containing nickel are
less corrosion resistant in these environments than
ferritic nickel-free alloys as shown in Fig. 22. Thus
ferritic 2¼Cr steels are standard in UK oil fired
plant, associated with a maximum steam temperature of
540°C, because of the steel's lower strength compared
with an austenitic. Even so, overheating can readily
cause accelerated rates of corrosion, exacerbated by
steam-side oxidation as described below.

3.2.3 Steam-side oxidation

In the past, steam-side oxidation has not needed to
receive such detailed attention as the fireside (or
indeed as the water-side, particularly in the
evaporator tubes). However, in the last 15 years two
major aspects of steam-side oxidation have required
extensive study, one for austenitic and one for
ferritic steels.

The first to come to the fore was the occurrence
of tube blockages in the final austenitic superheaters
of certain power stations. Because the gas temperature
is over 1000°C, once the steam flow is obstructed the
metal temperature rises rapidly and, as the tube is
still pressurised, failure by creep can occur within
hours. It was established that the debris was the
outer layer of the two-layered oxide which forms on the

Type 316 final superheaters. The morphology of the oxide is similar to that shown in Figure 4. The outer layer is magnetite (Fe_3O_4) with generally some haematite (Fe_2O_3), whilst the inner layer is a spinel containing iron, chromium and the other alloying elements. It was shown that loss of the outer layer did not affect the subsequent oxidation rate - i.e. rate control is in the inner layer and is approximately parabolic. Thus once the critical oxide thickness had grown so that it would spall, regrowth until a critical thickness was reestablished would take a much longer time than that to the first event.

The requirements were to be able to predict when the first and subsequent spalling events would occur. To do this a spalling criterion based on the energy required to produce the new surface which occurs during spalling was developed[5]. This energy comes from the thermal mismatch strains which occur when the oxide and its substrate metal are cooled from the oxide growth temperature, because of their different thermal expansion coefficients. The energy in a strained material is given by $\sigma\epsilon/2$ per unit volume, where ϵ is the strain produced by stress σ. In an oxide of thickness h, this can be expressed as $\frac{hE\epsilon^2}{2(1-\nu)}$ per cm², where ν is Poisson's ratio. This mismatch energy can then support the formation of a new interface if it is greater than 2γ where γ is the surface energy per unit area. This concept was shown to explain the compressive energy required to cause the observed spalling, and has been extended to other systems, such as described in Section 2.3.

The technological problem was not specifically that the oxide spalled, but that it did so in large fragments which could build up a blockage. Thus, techniques were examined which could affect the oxide formed and/or its spalling characteristics. For instance, cold work causes a chromium rich oxide to form, and changes in steam quality can affect the mismatch energy as it has been shown that the ratio of Fe_2O_3 to Fe_3O_4 critically affects the mismatch energy. It was also established that other austenitic steels did not oxidise to form such large oxide debris.

The concept was extended[5,10] to apply to both compressive and tensile stresses and an oxide failure map was derived - Fig. 23. This shows how thin oxides can withstand greater stresses than thicker ones, and should be compared with the experimental data shown in Fig. 14. It is also not necessary to consider only thermal mismatch stresses. Other sources of strain in the oxide would be equally effective, such as operational stresses, and the "system" strains which occur when a large multi-material structure is cooled.

The second steam-side oxidation phenomenon which has caused extensive technological problems, especially in the United States, is the formation of anomalously thick oxide layers on 2¼Cr superheaters, headers, valve chests, etc. In the US the thick oxide has been found to spall, and the oxide debris has been able to pass through to the turbine and cause extensive damage – Solid Particle Erosion[11]. In the UK such thick oxide has been found only on the bores of the final superheaters in some oil fired stations. In this case the oxide has not spalled, but has so adversely affected the heat transfer that the metal temperature has increased leading to accelerated corrosion and creep of the tube.

Examination of the oxide shows a multi-laminated scale – Fig. 24 – being a succession of the double oxide layer normally found on this material with an inner layer of iron chromium spinel and an outer layer of magnetite. The explanation of the UK behaviour is believed to be that the tubes was initially being overheated. Significant creep of the tubes was therefore occurring and the tubes were thus dilating. As a result the oxide could not relax back on to the metal and a new layer of oxide began to build up at the oxide-metal interface (the alternative could have been through scale cracking). This thicker oxide caused an extra insulating effect and hence the metal temperature to increase slightly. This increased the creep rate slightly and the process accelerated to rapid tube failure.

This mechanism requires significant creep deformation of the metal. It is difficult to see why heavy section components would oxidise by this mechanism, but laminated scales form in the US even on thick section components, and spall and cause turbine damage. The reasons for this difference in behaviour are not yet clear.

4.   CONCLUDING REMARKS

This paper has used a range of examples from the power generating industry, to illustrate some of the important corrosion mechanisms and the methods by which the industry deals with them.
   The importance of corrosion to nuclear operational safety has resulted in very large oxidation programmes to validate the behaviour of the materials. The understanding of the corrosion mechanisms in these environments has played an important part in establishing the confidence in the predictions. The well defined environmental conditions leads to readily achievable laboratory simulations.
   In fossil-fired boilers tube failures and excessive wastage rates result purely in economic penalties. What is acceptable as a "solution" has to be justifiable by economic assessments, and in this technology where access for repairs is not difficult, the cost-effective solution is often not that which gives the lowest corrosion rate. The presence of many aggressive impurities in a wide range of concentrations leads to more difficulty in simulating plant behaviour in the laboratory than for gas cooled reactors. However, such laboratory experiments are necessary to understand the corrosion mechanisms which can occur, and they have been respoonsible for some of the major advances in controlling fireside corrosion.

5.   ACKNOWLEDGEMENTS

This paper is published by permission of the Central Electricity Generating Board.

6.   REFERENCES

1.   T.N. RHYS-JONES: in 'Surface Stability' (ed. T.N. Rhys Jones), Chapter 1; 1989, London, Institute of Metals.
2.   P.C. ROWLANDS: J.C. GARRETT and A. WHITTAKER, CEGB Research, No. 15, November 1983, 3-12.
3.   P.C. ROWLANDS, J.C.P. GARRETT, L.A. POPPLE, A. WHITTAKER and A. HOAKSEY: Nuclear Energy, 1986, 25, 267-275.

4. H.E. EVANS: Mats. Sci. and Tech., 1988, $\underline{4}$, 415-420.
5. J. ARMITT, D.R. HOLMES, M.I. MANNING, D.B. MEADOWCROFT and E. METCALFE: "The Spalling of Steam-Grown Oxide from Superheater and Reheater Tube Steels", Report FP-686, Electric Power Research Institute, Palo Alto, California, 1978.
6. J.W. LAXTON, D.B. MEADOWCROFT, F. CLARKE, T. FLATLEY, C.W. KING and C.W. MORRIS: "The Control of Fireside Corrosion in Power Station Boilers (Third Edition, 1989)", Central Electricity Generating Board, London.
7. D.B. MEADOWCROFT: Materials Science and Engineering, $\underline{88}$, 1987, 313-320.
8. A.J.B. CUTLER, T. FLATLEY and K.A. HAY: CEGB Research, No. 8, October 1978, 12-26.
9. E.P. LATHAM, D.B. MEADOWCROFT and L. PINDER: Materials Science and Technology, to be published.
10. M.I. MANNING: in V. Guttmann and M. Merz (eds.) "Corrosion and Mechanical Stress at High Temperatures", Applied Science, London, 1981, 323-338.
11. J.F. QUILLIAM (ed): "Solid Particle Erosion of Utility Steam Turbines: 1985 Workshop", Report CS-4683, Electric Power Research Institute, Palo Alto, California.
12. J.E. FORREST and P.S. BELL in V. Guttmann and M. Merz (eds.) "Corrosion and Mechanical Stress at High Temperatures", Applied Science, London, 1981, 339-358.

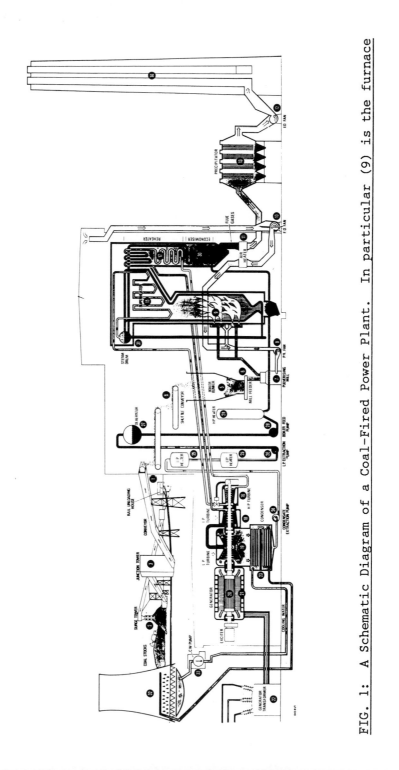

FIG. 1: A Schematic Diagram of a Coal-Fired Power Plant. In particular (9) is the furnace

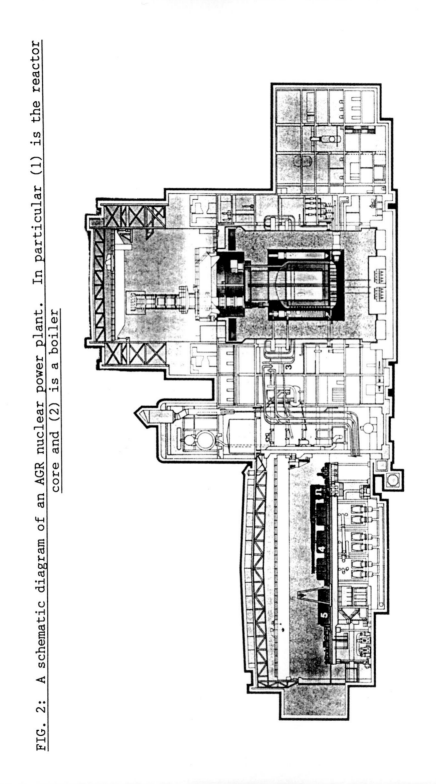

FIG. 2: A schematic diagram of an AGR nuclear power plant. In particular (1) is the reactor core and (2) is a boiler

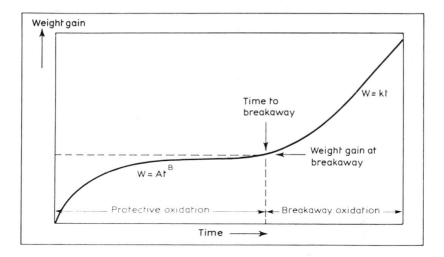

FIG. 3: The various parts of a weight gain-time dependence for an
oxidation process which breaks down from a protective to a non-
protective process. (after ref. (2))

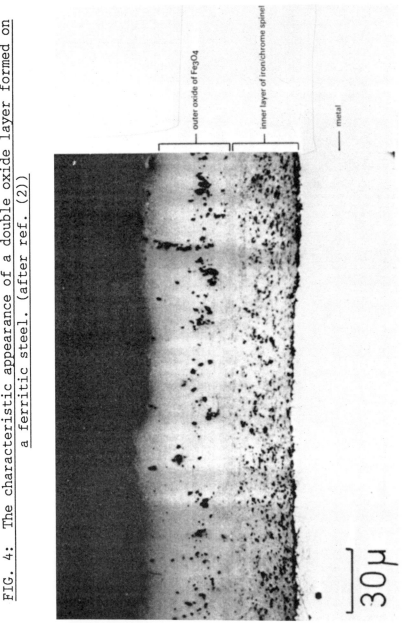

FIG. 4: The characteristic appearance of a double oxide layer formed on a ferritic steel. (after ref. (2))

outer oxide of Fe₃O₄

inner layer of iron/chrome spinel

metal

30μ

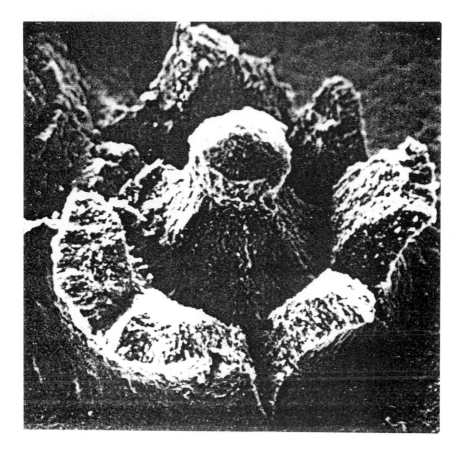

FIG. 5: A scanning electron micrograph of a breakaway oxide nodule on mild steel pushing through the protective oxide (after ref. (2)). The nodule is about 1 mm across

FIG. 6: The elongation of a bolted assembly with ∿3 cm diameter washers
– unoxidised on the left – after 4000 h at 500°C in high pressure CO_2
(after ref. (2))

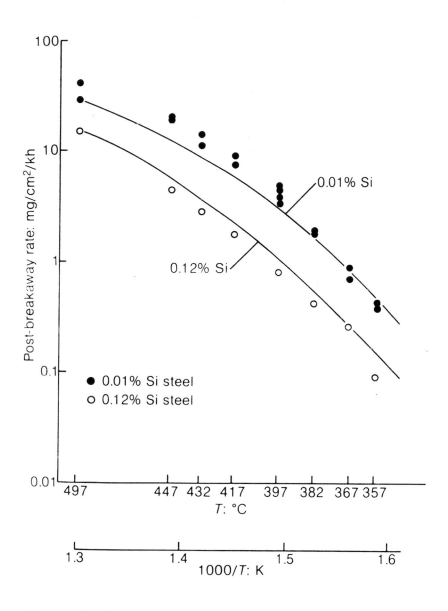

FIG. 7: The fit between the post breakaway oxidation rate equation for mild steel and two sub-sets of laboratory data (after ref. (3))

FIG. 8: Breakaway on a 9Cr sample (1 cm square coupon) occurring
preferentially at a corner (after ref. (3))

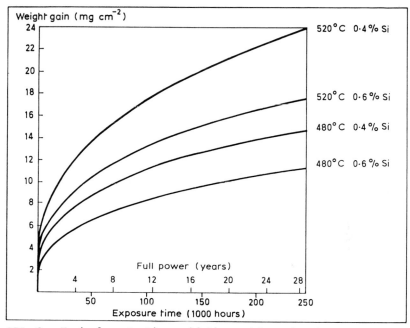

FIG. 9: Typical protective oxidation weight gain-time curves for 9Cr steels at two silicon contents (after ref. (2))

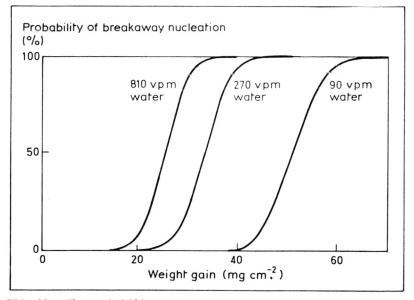

FIG. 10: The probability of breakaway for 9Cr steel at 640°C as a function of weight gain and gas water content (after ref. (2))

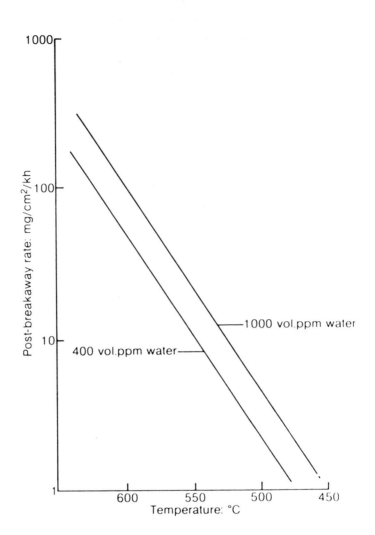

FIG. 11: The post breakaway rate for 9Cr steels as a function of
temperature and water content in the gas (after ref. (3))

30 μm

FIG. 12: An example of local pitting attack on 20Cr25Ni alloy after spalling of the protective oxide (after ref. (4))

FIG. 13: Possible spalling mechanisms caused by the oxide being in compression (after ref. (4))

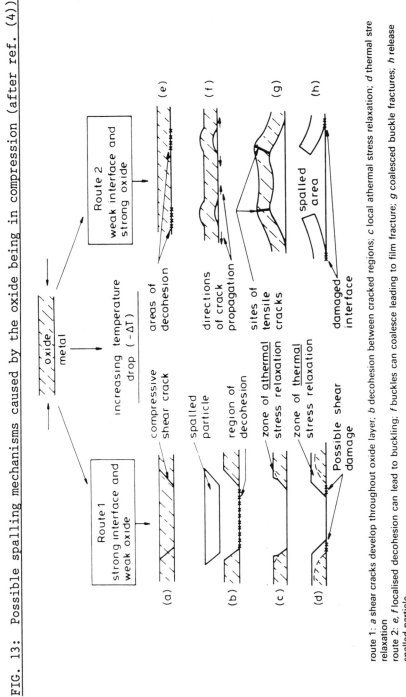

route 1: a shear cracks develop throughout oxide layer; b decohesion between cracked regions; c local athermal stress relaxation; d thermal stress relaxation

route 2: e, f localised decohesion can lead to buckling; f buckles can coalesce leading to film fracture; g coalesced buckle fractures; h release spalled particle

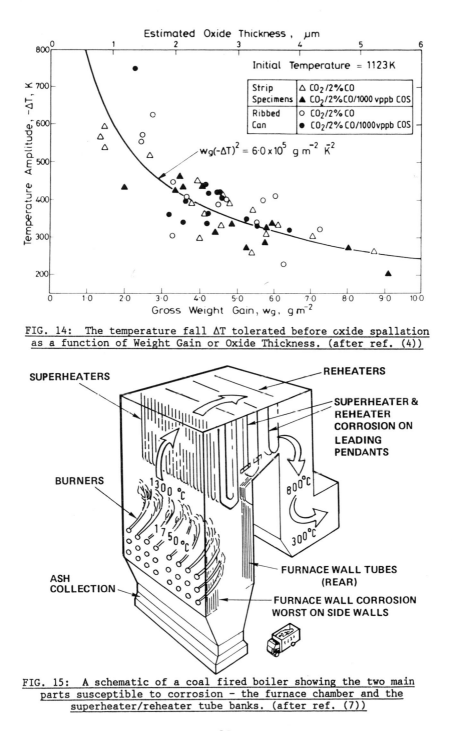

FIG. 14: The temperature fall ΔT tolerated before oxide spallation as a function of Weight Gain or Oxide Thickness. (after ref. (4))

FIG. 15: A schematic of a coal fired boiler showing the two main parts susceptible to corrosion – the furnace chamber and the superheater/reheater tube banks. (after ref. (7))

FIG. 16: A typical superheater tube bank in a modern boiler
(after ref. (8))

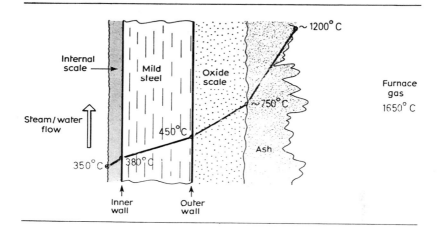

FIG. 17: The temperature distribution across a furnace wall tube
caused by the high heat flux. (after ref. (8))

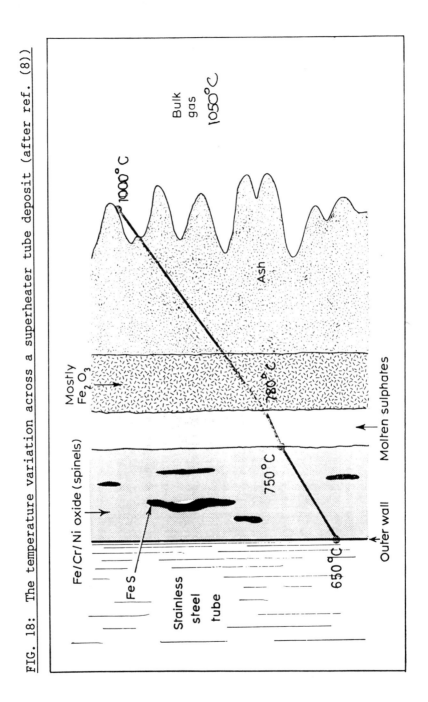

FIG. 18: The temperature variation across a superheater tube deposit (after ref. (8))

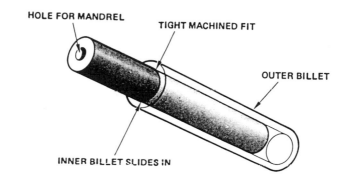

HOLE FOR MANDREL

TIGHT MACHINED FIT

OUTER BILLET

INNER BILLET SLIDES IN

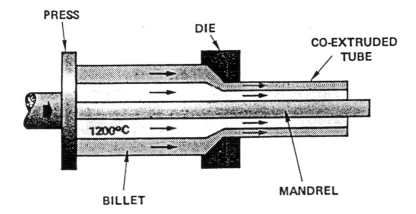

PRESS

DIE

CO-EXTRUDED TUBE

1200°C

BILLET

MANDREL

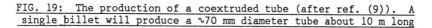

FIG. 19: The production of a coextruded tube (after ref. (9)). A single billet will produce a ∿70 mm diameter tube about 10 m long

89

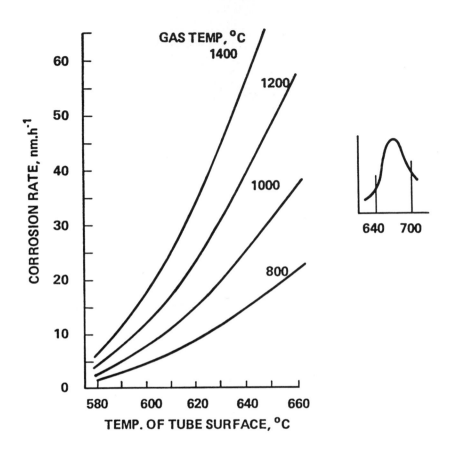

FIG. 20: The corrosion rate of austenitic steels in coal-fired boilers for a fixed chlorine content. The inset shows that from laboratory and probe tests the corrosion rate decreases at higher temperatures (after ref. (7))

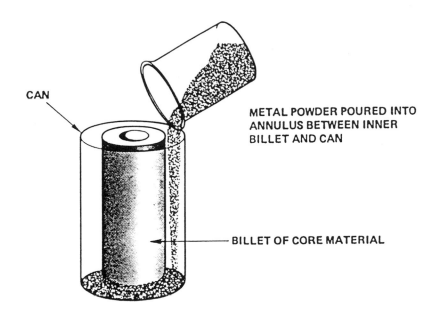

CAN

**METAL POWDER POURED INTO
ANNULUS BETWEEN INNER
BILLET AND CAN**

BILLET OF CORE MATERIAL

FIG. 21:   The powder metallurgy route used to produce billets for
 coextruded tube production with a 50Cr50Ni outer (after ref. (9))

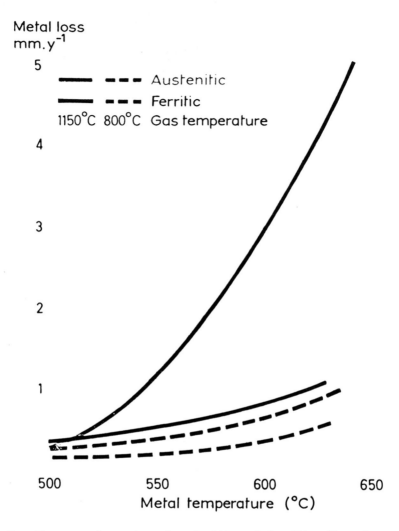

FIG. 22: The corrosion rates of austenitic and ferritic alloys in an oi
fired boiler (after ref. (7))

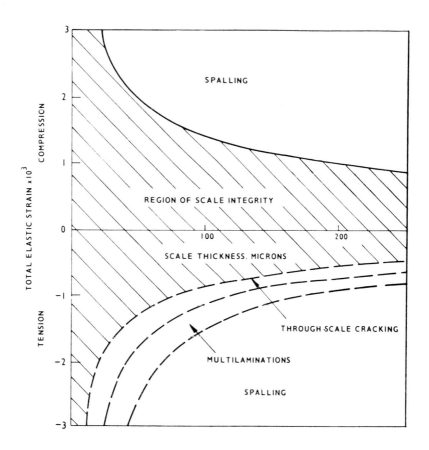

FIG. 23: An oxide failure map for tensile and compressive strains
(after ref. (10))

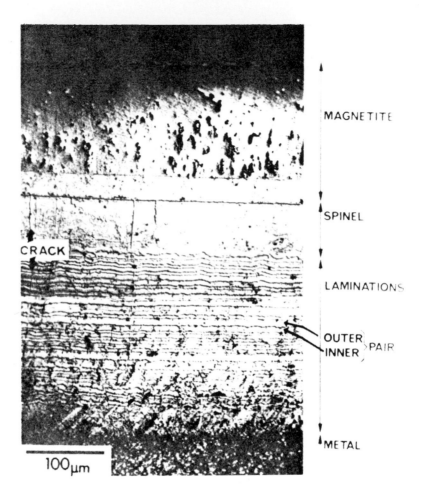

MAGNETITE

SPINEL

CRACK

LAMINATIONS

OUTER ⟩ PAIR
INNER

METAL

100 μm

FIG. 24: The anomalously thick laminated oxide which can form on low chromium ferritic steels (after ref. (12))

3:Testing methods in high temperature corrosion
J R NICHOLLS AND S R J SAUNDERS

1   INTRODUCTION

The performance of materials at high temperatures is dominated by their resistance to mechanical deformation and attack by the environment.   Thus, in attempting to select materials for service in these conditions it is important to have reliable means of predicting their behaviour, and whilst some information can be obtained by using the experience gained in the operation of existing plant, introduction of new materials or of different operating conditions requires data to be produced from laboratory testing procedures. High temperature corrosion can involve attack in the gas phase or the combined effects of hot gases and molten or solid deposits. As might be expected, if solid deposits can form, an increase in gas stream velocities would result in erosion rather than deposition and this will also addressed.   As will become evident the mechanical behaviour of oxide layers is an extremely important factor determining the overall performance in corrosion and erosion processes, so that the available methods of ascertaining the mechanical properties of thin layers will also be considered.

High Temperature Corrosion
Before describing the various test methods in detail it is appropriate to briefly discuss the fundamentals of corrosion processes in the various cases to be considered.

a)   Attack by hot gases
Classical thermodynamics can be used to determine the temperature and gas pressure at which the following reaction would occur:

$$M + X \rightarrow MX$$

where M is a metallic material and X is a component of the atmosphere.

At equilibrium for that reaction,

$$\Delta G^o_{MX} = \frac{RT}{2} \, \ln p(X)$$

where ΔG^o_{MX} is the standard free energy of formation of the compound and $p(X)$ is the partial pressure of the reactant in the gas atmosphere. Commonly X is oxygen but many industrially relevant environments contain halogenic (most frequently chlorides) sulphurous and carbonaceous species so that it is important to include a consideration of $p(Cl_2)$, $p(S_2)$ and the activity of carbon, a_c. Thus, frequently competing reactions occur with either oxide, sulphide, chloride or carbide formation.

Reaction with the environment is generally limited by ensuring that the corrosion product itself acts as a barrier between the atmosphere and the corroding material. The corrosion product is an effective barrier if rates of diffusion are low and if the layer is both chemically and mechanically stable in the environment. In this respect oxide layers are usually superior to most other corrosion product layers, and by choice of alloying additions it is frequently possible to ensure formation of layers of alumina, chromia or silica which generally have the best combination of properties as protective layers. A detailed consideration of the fundamental principles of high temperature oxidation and corrosion has been given in a number of excellent books and reviews to which the reader is referred to for more information[1,2,3,4,5].

b) Attack by molten salts

In some environments combustion of fuel results in the formation of sodium sulphate or, if vanadium is a constituent of the fuel, vanadium pentoxide. Both these species are volatile in hot combustion gas, but downstream of the combustion zone condensation as a molten phase can occur if components are present with temperatures lower than the dewpoint. In these circumstances a layer of molten salt deposits on the protective oxide film, and the stability of the film is then controlled by its chemical resistance to these molten salts[6].

c) Attack by solid deposits

Solid deposits can form on some components of industrial plant, usually associated with combustion processes particularly of coal, which produces relatively large amounts of ash. The deposit layer builds up on the oxide thus isolating it from the ambient atmosphere. In certain circumstances the environment at the deposit/oxide interface is altered and this can result, for example, in a build–up of sulphurous gases in the pores of the deposit which could lead to sulphide formation in conditions where only oxides would form in the absence of the deposit[7]. Similar effects have been noted if the oxide layer itself is porous and in CO_2/CO gas mixtures carbon formation has been observed in these porous scales[8].

d) Mechanical effects

The formation of a product layer on a metal or alloy usually results in a volume change, and for many oxide layers formed on high temperature alloys the oxide layer has approximately twice the molar volume of the metal. Depending upon whether scale formation occurs predominantly at the scale/gas interface (growth controlled by cation diffusion) or at the scale/alloy interface (anion diffusion) stresses may be imposed upon the scale as a consequence of the growth process. Simplistically, the latter process of growth controlled by anion transport would be expected to result in highly stressed scales. The process involved are complex, however, and for the present purposes it is not necessary to discuss this in detail; The salient processes involved in stress generation and relief have been reviewed elsewhere[9,10,11,12] to which reference should be made for an introduction to this topic.

Another important consideration is that frequently there is mismatch in thermal expansion coefficient between the scale and the substrate, so that if thermal cycles are imposed compressive stresses usually arise in the scale and this can also result in mechanical failure and loss of protection. Recently, methods have been devised to predict scale spalling using a fracture mechanics approach (13,14) or fracture strain/interfacial energy considerations[15].

2 MONITORING OXIDATION PROCESSES

The methods used to measure the rates of oxidation, ie. processes forming only oxide layers, are considered in this section but as will be evident many of the techniques are applicable to the other reactions and processes discussed in later sections of this chapter. In the first place consider the simplest process in which a test is carried out isothermally.

a) Isothermal testing

The purpose of tests of this type is to determine the rate at which the specimen is converted to oxide, and the most commonly used procedure is the gravimetric method in which the change in weight of the sample is recorded. Usually a weight gain corresponding to uptake of oxygen is observed, but if oxide spalling occurs this weight gain is reduced by the amount of metallic oxide lost, and in extreme cases net weight losses can be recorded. The simplest form of the experiment is to place the sample directly into the furnace withdrawing it periodically and measuring the change in weight. This experiment clearly imposes temperature cycles, which are discussed later, but in some cases particularly for thin adherent oxide layers, similar results are obtained with experiments carried out continuously. Figure 1 shows schematically a typical arrangement for conducting such experiments with continuous weighing of the sample using a microbalance. The best modern instruments have a sensitivity of $1 - 5$ μg with maximum weights of sample about 5 grams; heavier samples can be used but usually this involves some loss in sensitivity. To achieve this sensitivity it is important to ensure that the balance is isolated from vibration, although with currently available instruments this is achieved internally and bench-top models can be obtained. For a typical material the weight gain of 1 μg cm^{-2} corresponds to an oxide layer about 6 nm thick.

The results of these experiments are usually expressed as weight change versus time plots, (Figure 2) from which it is possible to derive the rate constant. In the simplest case for lattice diffusion controlled growth through a single phase adherent scale a parabolic relationship of the form

$$ w^2 - 2k_p' t - k_p t $$

is observed, where w is the weight gain, t is the time and k_p is known as the parabolic rate constant. Various other rate constants are used: k_t represents the rate of formation of the number of molecular units per square centimeter per second for an oxide 1 cm thick. Wagner[16], who carried out the initial theoretical analysis, defined a rational rate constant, k_r, expressed in equivalents of oxide (M_aO_b) per square centimeter per second. The relationship between the various constants is

$$ k_p((gO_2)^2 \ cm^{-4} s^{-1}) \ - \ \frac{2db^2 M_o^2}{NM_{M_aO_b}} k_t \ - \ \frac{dbM_o^2}{M_{M_aO_b}} k_r $$

$$ k_p'(cm^2 \ s^{-1}) \ - \ \tfrac{1}{2} \frac{M_{M_aO_b}^2}{M_o^2 d^2} k_p((gO_2)^2 \ cm^{-4} s^{-1}) $$

where d is the density, N is Avagadro's number, M_O is the atomic weight of oxygen, $M_{M_a O_b}$ is the molecular weight of the oxide $M_a O_b$.

Measurement of pressure changes in a closed system has also been used to monitor rates of oxidation. This type of method is only applicable in pure gases since consumption of one component of a mixture would alter the composition and consequently may affect rates of attack. This method is not widely used, although using a differential manometer Campell and Thomas[17] were able to detect film thickness corresponding to monolayer formation. A variant of this type of method involves measurement of changes in volume as the oxidising species is consumed. This does, of course involve a change of pressure during the course of the reaction but providing the sizes of the reservoir and adsorption chamber are carefully chosen this is not a serious limitation.

Optical methods have been used very effectively to measure oxide film thickness and probably the most important of these is ellipsometry. The procedure consists of reflecting polarised light from the oxidised metal surface and measuring the state of polarisation. The changes in polarisation observed depend on the angle of incidence and the optical properties (refractive index) of the film. Polarised light can be resolved into two components, parallel and perpendicular to the plane of incidence, and these two components suffer different phase retardations, Δ, and reductions in amplitude, tan ψ, on being reflected from a surface. Measurements of Δ and tan ψ for a film-free metal are made and the optical constants of the metal can then be calculated. Oxidation of the surface takes place and values of Δ and tan ψ are obtained as a function of time, so that the film thickness and refractive index can then be computed. A complete description of this and various other methods is unfortunately beyond the scope of the present chapter but the reader is referred to Kubachewski and Hopkins[2] for full details.

Oxidation rates also have been determined by measurement of the natural frequency of a freely suspended specimen by Hancock and co-workers[18,19,20], and as will be shown later in the section of mechanical stability of oxides, cracking and spalling of the scale can also be detected. The characteristic frequency of the first mode of vibration, f_O, is related to the dimension and the elastic modulus, E, of the rod as follows:

$$f_o^2 = \frac{(4.73)^4 E a^4}{16 \pi l^3 m_i}$$

where a, l and m_i are the radius, length and initial mass of the rod.

If the rod now forms an outer oxide layer the relationship changes.

$$f^2 = \frac{(4.73)^4}{16\pi l^3 m_c} [E_m c^4 + E_o(b^4 - c^4)]$$

where E_m and E_o are the elastic modulie of the metal and oxide, m_c is total mass of the composite rod, c is the radius of metal and b is the radius of the composite.

Thus monitoring frequency changes can give information about the mass changes and the following simplified expression has been derived:

$$\frac{f^2 - f_o^2}{f_o^2} = \Delta \left[\frac{2\beta M_m}{M_o - M_m} - \frac{M_o - M_m}{M_o - M_m} \right]$$

where Δ is defined by $m_c = m_i(1+\Delta)$
   f is frequency of the oxidised rod
   M_m, ρ_m, E_m are the atomic weight, density and elastic
            modulus of the metal
   M_o, ρ_o, E_o refer to an oxide

$$\beta = \frac{E_o \rho_m M_o}{E_m \rho_o M_m}$$

b) Cyclic Oxidation

The superimposition of thermal cycles in an oxidation experiment, cyclic oxidation, is used to encourage scale failure due to spallation[21,22,23]. Hence cyclic oxidation tests are used to monitor both scale adherence and the ability of a material to successfully repair after repetative scale failure.

The materials performance is generally monitored thermogravimetrically, either continuously[21,22] or in a discontinuous manner[22,23], although other monitoring methods have been used. For example, measurement of the specimen's resonant frequency during thermal cycling has been successfully used to detect[19] the onset of scale failure, before gross spallation, and hence weight loss, was observed. Figure 3 illustrates the typical data produced by this form of test. Curve (a) represent good

100

performance with little scale failure; curve (b) is intermediate behaviour that is initially protective, but after a limited test duration becomes non–protective, while curve (c) poor behaviour with scale spallation occurring from the onset of the test. Curve (b) is characteristic of materials that have a limited reserve of stable scale forming elements, such as chromium, aluminium and silicon, so that after repeated scale failure and repair the morphology and composition of the scale changes, possibly forming less protective spinels for example. These less protective scales spall more readily and rapid weight losses are recorded[23].

With this general behaviour in mind, one may look to the successful design of a cyclic oxidation experiment. The critical parameters include:–

i) Temperature change (ΔT) and cooling rate. These parameters determine the degree of compressive stress induced on cooling. If the stress exceeds some critical value (determined from fracture mechanics considerations[13,14]) then scale spallation can result. Two modes of scale failure have been observed[24,25], the first requires the generation of through thickness shear cracks and then decohesion[24], while the second results from loss of adhesion, scale buckling and then failure[25]. This second mode of failure can only reasonably apply to thin scales ($< \frac{1}{2} \mu m$ thick) while the first mode has much wider applicability. For the first failure route, the onset of spalling is given by

$$\Delta T \ (\text{spall}) = \left[\frac{\gamma_F}{f d E_0 (\alpha_m - \alpha_0)^2 (1 - \nu_0)} \right]^{\frac{1}{2}}$$

where γ_F is the energy necessary per unit area to fracture the metal oxide interface, f is the fraction of stored energy used in the fracture process (~ 1), d is the oxide thickness, E_0 and ν_0 are the elastic modulus and Poisson's ratio for the oxide and α_m and α_0 are the expansion coefficients of the metal and oxide respectively. This critical temperature drop is plotted against oxide thickness in Figure 4, for a chromia scale ($E_{ox} = 260 \ \text{GNm}^{-2}$, $\nu = 0.3$, $\alpha_m - \alpha_0 = 9.6 \times 10^{-4} \ \text{K}^{-1}$,) for various values of γ_F. At 700 °C, cohesive failure within the oxide results in a fracture energy between 5 and 10 Jm^{-2} for chromia scales. At lower temperatures lower fracture energies are to be expected due to reduced scale plasticity, hence from Figure 4 temperature drops in excess of 400 °C should descriminate between good and poor scale adhesion for oxides in the range 1 – 3 μm thick.

ii) Time at temperature directly influences scale thickness and hence the number of cycles before an initial oxide spall is generated and also the ability of the scale to repair after the thermal cycle. A range of cycling rates has been reported in the literature, varying from 20 per hour to 1 every 20 hours depending on the service cycle expected. A cycle time of one every 2 hours proves a useful compromise. It is sufficiently descriminatory but still permits sufficient scale growth and repair between cycles.

As stated earlier, both continuous and discontinuous monitoring of cyclic oxidation have been reported in the literature. The simplest test methods involve mechanical cycling of specimens into and out of a furnace at temperature. Convective cooling or forced air cooling has been used depending on the desired cooling rate. Specimen mass change is measured either at each cooling cycle or after a known number of cooling cycles depending on cycle frequency.

A somewhat more sophisticated test procedure, which has been widely used[23,26], involves cycling a carousel of specimens from a hot combustion gas stream into a cooling air blast. Figure 5 illustrates a typical burner rig facility[26] that is used for this form of cyclic oxidation testing. Again mass change for each specimen is monitored intermittently during the cooling cycle. Because this approach uses combustion gas products during the heating cycle, samples under test may exhibit gas phase corrosion instead of pure oxidation depending on the purity of the fuel and the material under test. Gas phase, molten salt and solid deposit corrosion test procedures will be discussed later in this chapter.

Continuous monitoring during cycle oxidation tests has been reported using thermogravimetric[21,22] and resonant frequency[19] monitoring techniques. Both methods benefit from the use of programmable temperature controllers to effect the thermal cycle, however, this limits the cooling and heating rates that can reasonably be achieved.

The resonant frequency technique is particularly suited to the determination of the first crack in the oxide as oxide cracking modified the effective stiffness of the composite system, which can be detected as an instantaneous drop in the resonant frequency. However continued spall formation following the initial cracking will produce little further change in frequency. Conversely, thermogravimetric methods are unlikely to detect early crack formation unless rapid oxidation occurs over a sufficiently large area of the surface. This method, however, is particularly suited to the monitoring of continued spall formation during cooling.

c) Mechanistic studies
Various experimental procedures have been devised specifically to

obtain information about the mechanism of oxidation and should be included in any discussion of test methods. When considering mechanisms of oxide growth it is important to know the transport path (lattice or grain boundary diffusion), where oxide growth is occurring, ie. anion or cation movement, the diffusion coefficients of the more mobile ions and the equilibrium defect concentration. Tracer and marker techniques can be used to determine the nature of mobile species and transport paths and the Rosenburg method, using interrupted kinetics experiments, gives information about diffusion coefficients and defect concentrations.

i) Marker methods
If inert material, platinum or a stable oxide, is placed on the surface of unoxidised metal its final position in the oxide scale will give information about the transport processes that have occurred as is illustrated in Figure 6. If the scale is pore-free the final position of the marker ideally reflects the transport processes that have taken place. However, the presence of pores can result in the marker being displaced. For example, in cases where an outer dense layer forms over an inner porous region, the marker is frequently found at the interface between the porous and dense layers even though cation lattice diffusion is faster than anion diffusion. In this case it is believed that molecular oxygen has penetrated the outer layer as the result of cracking of the scale or of the development of microchannels due to build-up of stress. So that the final position of the marker reflects movement of material but not solid-state diffusion processes. It is also suggested that with large markers, the marker itself can partially obstruct the diffusion process and some plastic flow can occur around the marker.

ii) Tracer methods
This involves essentially the same principle as was used for the case of markers but a radioactive species of one of the reactants in the oxidation process is used. This can be done by either depositing a thin layer of radioactive metal on the surface of the specimen or by carrying out the experiments with radioactive oxidant at some stage in the experiment; ^{18}O is commonly used in oxidation experiments. Atkinson et al[27] have calculated idealised profiles for various limiting cases which are reproduced in Figure 7, where it can be seen that for the case of a scale, MO, growing by vacancy diffusion the profile depends on the effective charge, α, of that vacancy. Where grain boundary diffusion occurs it has been assumed that there is no exchange of metal between the grain boundary and the oxide lattice and thus the cations are immobilised. In fact, some exchange does occur and this would

have the effect of broadening the profile.

In most practical cases some combination of the limiting cases referred to above is usually found so that the resultant profile will be an appropriate mixture of the simple cases.

iii) Rosenburg method[28]

The rate of oxidation reactions is expressed in terms of the rate constant, k_p', which can be shown to equate to

$$\Omega D_d (C_o - C_1)$$

where Ω is the increase in volume of a defect upon completion of the reaction, D_d, is the diffusion coefficient of the defect and C_o and C_1 are, respectively, the equilibrium concentrations of defects at the originating interface and the interface where the reaction is completed. Since C_1 is generally small compared with C_o and Ω is known, measurement of the parabolic rate constant gives the product $D_d.C_o$ directly. However, the individual values (D_d and C_o) are required but these are normally difficult to obtain and are only valid if the structural characteristics (grain size etc) of various samples are similar. Rosenburg[28] has shown that it is possible to obtain this information relatively simply by carrying out interrupted oxidation experiments in which the oxidant is removed from the system while maintaining the sample at temperature and holding for sufficient time to allow equilibration to take place, and then readmitting the oxidant. Figure 8 shows idealised concentration profiles of defects during reaction and interruption.

An analysis of the reaction immediately after interruption shows that Δ, the number of defects crossing unit area at x=X in time t is

$$\Delta = Bt^{\frac{1}{2}} \qquad (t \ll X^2/D)$$

where $B = 1.12 \, D^{\frac{1}{2}} C_o$
and later

$$\Delta = Q + Rt \qquad (t > X^2/2D)$$

where $Q = XC_o/3$ and $R = DC_o/X$

so that the following expressions for C_o and D can be obtained

$$C_o = \frac{0.8\,B^2}{XR} = 3Q/X$$

$$D = 1.25 \left[\frac{RX}{B} \right]^2$$

where B, Q and R can all be measured.

This analysis is only valid for scales controlled by lattice diffusion processes and has found use only in the more acedamic studies of scale formation[29].

d) Post test evaluation

Continuous assessment of the rate of reaction as outlined above generally gives kinetic information which can be analysed in various ways to provide, either information on the mechanism of the process, or a sound basis for predicting long-term behaviour. Breakaway oxidation, which involves a transition from protective (parabolic) kinetics to more rapid non-protective (linear) kinetics, will be considered in the section on mechanical stability of scales since the main reason for this phenomena is usually associated with scale rupture associated with stress in the oxide layer attaining some critical value. In this case, of course, extrapolation of the protective regime can result in seriously misleading conclusions should breakaway subsequently occur. For this reason, amongst others therefore, it is important to examine the scale after completion of short-term tests to ascertain its structure and the morphology of attack. The preferred method of carrying out such an investigation is to prepare a polished metallurgical cross-section or fracture section for examination in the optical or scanning electron microscope. For thin films, separation techniques involving "peel-off" methods or dissolution of the metal substrate allow the film to be examined in the transmission electron microscope. The experimental methods involved in specimen preparation and analysis have all been detailed in a previous monograph on "Microstructural Characterisation"[30].

All the kinetic methods of monitoring oxidation only give average information about the behaviour of the specimen, and this can frequently give misleading information. Corrosion processes are generally not uniform over the surface of a specimen so that while 90% of the surface may be covered with a thin protective film, it is perfectly possible for the remainder of the sample to have undergone the transition to breakaway corrosion. In these circumstances it is often difficult to adequately describe the thinning of a metallic sample and to develop predictive models. Nicholls and Hancock[31,32] have emphasised the use of statistical methods in assessing corrosion processes, and adopting this method can help to overcome some of the difficulties referred to above. In essence many measurement of the thickness of an oxide layer, remaining metal and/or internal penetration are made at random around the circumference of the sample and the values can be plotted, assuming that the data are normally distributed, on probability paper. This plot usually results in a straight line from which the probability of corrosion exceeding a given thickness can be determined. Figure 9 shows a probability plot for the corrosion of IN 939 at 700 °C that was obtained using the method outline above. The curve has two linear regions, the lower values result from oxidation while the higher values are due to a low temperature form of hot-salt corrosion. Separation of this bimodal distribution into component models demonstrates that each component is normally distributed[32].

In many corrosion processes it is the worst-case scenario that governs the lifetime of a component; for example failure of a boiler tube is governed by the deepest pit that is present. Normally this is very difficult to predict, but some guidance can be given by consideration of extreme value statistics, in which only maximum values of metal loss are considered. In this case the data are not normally distributed but have a skewed distribution which results in a curve when plotted using normal probability paper. An example, is given by Nicholls and Hancock[31] of the maximum observed pit depth on a sample being 99 μm but by use of the procedure outlined above they were able to calculate that an extreme value of 114 μm could be expected.

3. MECHANICAL FAILURE OF OXIDE SCALES

In the previous section on monitoring oxidation processes the importance of maintaining a protective oxide scale has been highlighted. Under these conditions oxidation rates are asymptotic (usually parabolic, but this depends on the oxidation temperature). The onset of linear kinetics therefore implies that scale failure has occurred (breakaway oxidation) and it becomes difficult to predict the rate of material loss and hence component life under these conditions. This section examines methods used to monitor scale mechanical properties and in particular the onset of scale failure.

It has already been shown that by careful design of a cyclic oxidation experiment the susceptability of a material to spall formation can be assessed. Hence cyclic oxidation tests provide a qualitative method of assessing scale adherence. To quantify scale adherence, and the onset of spalling, controlled cooling experiments are necessary. Continuous monitoring permits the onset of first fracture to be detected such that from a knowledge of the temperature drop the fracture energy for scale spallation can be calculated. This approach has been used by Evans and Lobb[33] to evaluate the interfacial fracture energy of chromia–rich scales formed on a 20Cr/25Ni/Nb–stabilised stainless steel after oxidation in $CO_2/2\%CO$ at 930 °C. A value of 5.8 Jm^{-2} was obtained.

Recent interest in predicting scale failure has centred around the use of fracture mechanics[13,14,34] to predict the onset of scale failure. In simplistic terms, scale failure occurs when the *in situ* stress exceeds some critical value (σ_c). The critical stress can be evaluated using simple fracture mechanics considerations[13,14], whether through thickness cracking (tensile failure) or scale spallation (compressive failure) results. For the case of tensile failure, the critical stress is given by

$$\sigma_c = K_{IC} . (\pi \bar{a})^{-\frac{1}{2}}$$

where K_{IC} is the fracture toughness and $\bar{a}$ is a measure of the defect size present in the oxide. K_{IC} is related to the energy of fracture by $K_{IC}^2 = 2 E_o \gamma_o$ where E_o is the oxide elastic modulus and γ_o is the energy required to form a unit area of new oxide surface. Hence fracture mechanics analysis[13,14,34] and the assessment of fracture strain energies[15] are alternative approaches to solving the same problem.

The *in situ* stress acting at a defect, within the oxide or at the oxide/metal interface, is the summation of all stresses acting on the defect. This stress will include growth stresses (σ_r), thermally induced stresses (σ_{th}) and directly applied stresses (σ_{ap}) which may be tensile, compressive or bending. Hence to quantify oxide fracture, the state of stress, fracture toughness and defect distributions present within the oxide or at the metal/oxide interface (depending on the mode of failure) must be capable of being assessed. This section examines test methods for evaluating the state of stress, fracture stress, fracture toughness and defect distributions present within the oxide scales.

a) **Internal stress measurements**
Two methods have been widely used to determine the internal or growth stresses present within oxide scales, these are X–ray stress analysis and methods based on the deflection of a thin beam open to oxidation on one surface only.

X-ray stress measurement techniques are usually evaluated at room temperature, although some facilities are available with a high temperature capability[35]. Hence the internal stress determinations include both growth stresses and thermally induced stresses during the cooling cycle.

The X-ray determination of stresses is based on the measurement of lattice strains in a set of lattice planes of known orientation and then correlating these lattice strains with the strain (ϵ) calculated from elastic theory[36]. This approach is known as the $\sin^2 \psi$ method, for which at the surface under plane strain conditions the following equation applies:-

$$\frac{da}{a} = \epsilon_{\varphi,\psi} = \left[\frac{1 + \nu}{E}\right] \sin^2\psi \ (\sigma_1 \ \cos^2\varphi + \sigma_2 \ \sin^2\varphi) - \frac{\nu}{E} \ (\sigma_1 + \sigma_2)$$

where σ_φ and ϵ_φ relate to the stress and strain measure in the direction φ of the specimen surface relative to the principal stress direction (σ_1). E and ν have their usual notation. By measuring da/a at selected values of φ a family of simultaneous equations can be derived which allows the evaluation of E, ν, σ_1 and σ_2 assuming that the surface is isotropic and hence that the material constants do not change with orientation.

The determination of internal and growth stresses by direct deflection measurements on a thin metal beam at temperature[37] has been widely used by Huntz and co-workers[38,39]. Figure 10 illustrates schematically this method of measurement. The thin metallic strip under test is only permitted to oxidise on one side by coating the second, usually with silica. Oxidation causes the strip to bend with the degree of deflection a measure of the stress generated within the oxide as defined by:

$$\sigma_o = \frac{E_m x_m^2 D}{3L^2 x_o}$$

where D is the beam deflection, L is the beam length, x_m and x_o are thickness of the metal and oxide respectively and E_m is the elastic modulus of the metal[38]. This measured stress is only a part of the total stress due to oxidation, as a residual stress component remains after beam deflection due to the tensile elastic strains generated within the metal beam[39]. When used isothermally at temperature this technique permits direct

108

measurement of the oxide growth stresses. Other methods based on the oxidation of foil specimens on one side have been used to evaluate growth stresses and these have been reviewed in reference 10.

Figure 11 is a typical example of the results obtained using the bending beam method for an Fe−45 Ni−25 Cr alloy oxidised at 1000 °C, with or without yttrium implantation[39]. The compressive stresses generated in the oxide reach a maximum (60−70 MPa) during early scale formation, reducing to a more stable value of 15−20 MPa as the oxide thickens. The implantation of yttrium can be seen to reduce the initial peak stress level for this alloy and hence the likelihood that the compressive growth stress is able to cause early scale failure by spall formation.

This technique could also be used to investigate the onset of spalling in controlled cooling experiments. Isothermal oxidation to a point of maximum growth stress, followed by controlled cooling would encourage oxide failure by spall formation. Loss of oxide could be directly detected and hence ΔT measured. Furthermore deflection of the beam during the cooling cycle would permit the effect of differential expansion coefficients to be evaluated and allow direct measurement of the imposed stress prior to failure. These controlled cooling experiments have not to the authors' knowledge been undertaken to date, but would greatly help in the quantification of oxide failure during thermal cycling.

b) Detection of scale failure

Three methods have been used to−date to detect scale cracking both during thermal cycling and under conditions of applied load. As discussed in the section on cyclic oxidation measurements controlled cooling experiments can reveal the critical temperature drops necessary to cause through scale cracking (increased oxidation rates are observed) or scale spallation (weight losses are recorded)[21,40,41,42]. However, this method is restricted to metal/oxide systems where weight changes induced by scale cracking can be readily detected. The vibration technique (as decribed in section 2a) can also be used to detect the onset of scale cracking[18,20,43]. It is very sensitive to crack formation within the oxide which results in a reduction of the stiffness of the composite system and hence a discontinuity in resonant frequency is observed (see Figure 13) when the oxide scale cracks. Continued cracking would result in a continual decrease in the recorded resonant frequency. Should oxide repair occur then a frequency transition is observed with the frequency recorded after repair approaching that observed before the cracking event.

The vibration technique offers the following advantages over thermogravimetric methods:

    i) it is very sensitive to the first crack in the oxide and can be used to monitor both continual cracking and oxide repair;

    ii) measurements taken in conjuction with thermogravimetry permit the oxide modulus to be evaluated *in situ*;

    iii) cracking can be detected both during thermal cycling and also by the application of a superimposed tensile stress.

However, this technique is limited to simple rod–shaped specimen geometries, requires care in preparing and setting up the specimens and must be monitored manually.

More recently acoustic emission has replaced the vibration technique as a route to measure in situ early scale failure. This technique is not limited to particular specimen geometries, has similar sensitivities to scale cracking and can be monitored automatically. However, it is unable to determine the elastic properties of the oxide *in situ*. This short–coming could be overcome by the use of acoustic microscopy which is a relatively new technique where an acoustic signal is focussed onto a sample using water as a coupling medium. Illett[44] has demonstrated that in the particular case of a thin cracked oxide partially detached from the substrate interference fringes form by Rayleigh scattering from which it was possible to calculate the elastic modulus. All other methods of obtaining these measurements give average values but this technique allows measurement of the modulus of an oxide scale of only a few μm^2, and interestingly significant variations in values were noted from area to area. Acoustic emission has found wide usage recently in monitoring cracking during thermal cycling[45,46], superimposed stress[47] and during isothermal oxide growth[48]. Figure 12 shows the typical coupling arrangement of an acoustic tranducer to a specimen through the use of an acoustic waveguide. Figure 12 specifically relates to monitoring acoustic events during the cracking of oxides under applied tensile loads[47], however, similar geometries can be used to monitor scale failure under cyclic conditions or during isothermal oxidation by combining acoustic emission with thermogravimetric studies[48].

Figure 13 compares schematically the results obtainable from thermogravimetry, acoustic emission and the vibration technique, when an oxidised specimen spontaneously enters a breakaway oxidation condition. Initially during parabolic growth (a) the weight gain and resonant frequency signal increase continuously and parabolically as the scale thickens. No acoustic events are recorded during this period. At the onset of breakaway an increase in the number of acoustic events is observed as the scale begins to crack (and spall). The resonant frequency of the composite (metal + oxide) at this stage decreases as the contribution of the oxide to the composite stiffness is lost (region b). Following breakaway a region of linear oxidation is observed with the scale continually cracking and repairing (region c). During this region of non–protective growth a continual, high level of acoustic events is recorded as the scale continues to crack. The resonant frequency response in this region is discontinuous where both increases and decreases in resonant frequency are observed; this behaviour results from a trade between loss of stiffness due to scale cracking and increased oxide thickness due to the linear growth kinetics. Finally should scale repair occur and protective parabolic kinetics be reestablished then a reduction in the number of acoustic events is observed with the specimen resonant frequency stabilising at some higher frequency than that measured before the onset of breakaway.

c) Measurement of the macro defects (cracks voids and pores) present in an oxide scale

As discussed earlier in this section, scale failure results when the *in situ* stress exceeds some critical value necessary to propogate a defect. Hence from a knowledge of the stress to cause fracture, the growth stress and the defect distribution in the scale, a measure of the scale fracture toughness and scale fracture energy can be determined. Until recently no method for evaluating scale defect distributions was available, but a recent paper by Hancock and Nicholls[14] has proposed a method for assessing the composite defect size in an oxide containing multiple voids and pores. The analysis follows guidelines proposed for the assessment of defects in welds, and requires that each defect is idealised and compared with its neighbours to assess if any interaction occurs. Interacting defects are considered as a single defect and the assessment procedure continues, until a single composite equivalent defect results. Using this defect size values of fracture toughness can be evaluated. Evidently, this process requires detailed metallographic analysis. The ability to quantify the defect volume fraction and distribution using image analysis techniques would greatly aid in the assessment of the defect distribution within the oxide scales and encourage a wider application of this analysis method. This approach has, however, been applied to a number of scale containing active elements and has clearly demonstrated that a major role of active element additions is to reduce the macro defects present within the oxide scales[34] with its consequent benefit on scale adherence and spalling resistance.

d) Qualitative methods of assessing scale adherence

The foregoing indicates that detailed analysis of stresses generated during oxidation and the defects present within the oxide are necessary to fully quantify scale failure and the oxide fracture toughness. However, often a more simple qualitative method of evaluating oxide scale adhesion is required. Three methods have been used to qualitatively assess adhesion. The most widely used method is an assessment of cyclic oxidation resistance (see section 2b). In addition to this direct separation tests[49,50] and scratch test methods[51] have been applied to oxide scales.

The direct separation test is illustrated schematically in Figure 14[49]. In this test a pull wire is bonded to the oxide surface and the load to fracture the oxide is measured. Failure may occur at the metal-oxide interface (adhesive failure) or within the scale (cohesive failure). This method is only really suitable for weakly bonded systems as the adhesive bond must be superior to the interfacial adhesion if realistic measurements are to be taken.

To overcome this bonding problem Nicholson (52) allowed a stainless steel gauze to be embedded in a scale as it forms on steel substrate. The gauze was attached to a steel rod to pull the scale at the test temperature. However, in spite of the simplicity of this technique, it suffers from several drawbacks. It is obvious that this method is limited to thick scales only, and no account was taken of edge effects. In addition, the presence of the stainless steel mesh probably influenced the growth rate and the measured adhesion strength.

A hot tensile test was used[53] to estimate the rupture strength of scales formed on mild steel at elevated temperatures. In this work tensile a test specimen was cut transversely at its centre and allowed to oxidise to provide the required joint between the two surfaces. A similar method was used to estimate the adhesion strength of scales formed on iron-base alloys[54]. Two pieces of metals of different diameters were placed on each other in intimate contact, ie. the interface remains relatively oxide-free. The oxide then grew on the outer sections of the specimens, welding the two halves together. In both of these methods tensile tests are used to assess the oxide mechanical properties at an appropriate temperature.

4 MIXED OXIDANT TESTS

Many industrial applications involve exposure of high temperature components to complex gas mixtures, and it is necessary to devise relevant laboratory test procedures. In designing the test it is important to consider whether the atmosphere being simulated is equilibrated or unequilibrated, and, also in context of the laboratory environment, whether adequate safety precautions exist since frequently the gas mixtures to be used contain toxic or combustible components. In this section consideration will be given to factors controlling the composition of the atmosphere and the experimental arrangements required to monitor gas composition, including any modification to experimental arrangements necessary to allow safe operation and introduction of samples.

a) Control of Gas Composition
Normally equilibrated gas mixtures are required, therefore, it is common practice to assume that the main oxidising (in the chemical sense) species control reaction rates, so that the partial pressures of oxygen (pO_2), sulphur (pS_2), halogenic gases (pCl_2) and the activity of carbon (a_c) are controlled by establishing relevant gas equilibria, for example,

H_2/H_2O	for	$p(O_2)$
CO/CO_2	for	$p(O_2)$ and a_c
H_2/H_2S	for	$p(S_2)$
SO_2/O_2	for	$p(S_2)$ and $p(O_2)$
CH_4/H_2	for	a_c
H_2/HCl	for	$p(Cl_2)$

It is relatively straightforward to carry out calculations to determine the partial pressure in these simple binary gas mixtures, but in multi−components atmospheres calculation of the equilibrated gas composition require the use of iterative routines such as the MTDATA in use at NPL[55]. The composition of the gas used in the test will, of course, control the type of scale that forms. For example, in an atmosphere that contains both oxygen and sulphur competition exists between oxide and sulphide formation. It is convenient to use the phase stability diagram of the relevant metal−oxygen−sulphur system to determine which phase would form, as shown in Figure 15 for the Cr−O−S system where provided the gas composition lies within the Cr_2O_3 phase field protective oxide formation is predicted. Usually, however, at low $p(O_2)$ values gas compositions must be more oxidising (by about 2−3 orders of magnitude) than the equilibrium values predict because of kinetic effects.[56]
    The choice of gas mixture is usually determined by a direct simulation (for industrial atmospheres or environments containing high partial pressures of reactive species) or by using a gas with an equivalent partial pressure of active species for low pO_2, pS_2, etc. Very recently, however, Kofstad[57] has questioned this latter approach, since he considers that the specific components of a gas mixture can individually control scale defect concentrations or microstructure thereby altering rates of attack. For example, it is known that H_2/H_2O of CO/CO_2 mixtures of equivalent $p(O_2)$ yield different rates of attack. Representative testing, therefore, should

be carried out with exact simulations of the relevant gas mixture were possible.

Gas mixtures are usually obtained by mixing at room temperature, or by using pre-mixed bottled gases, and since the equilibrium gas composition will change with temperature it is important to ensure that the gas has been heated to the test temperature before contacting the sample. In flowing gas streams, which are usually used to avoid depletion effects of active species (often present in only small quantities), it may be necessary to preheat the gas by passing it over a heat exchange system or in a serpentine path through the hot-zone of the furnace before contacting the samples. Some gas mixtures are slow to equilibrate so that use of catalysts is necessary, and a good example of this is the SO_2/SO_3 equilibrium which is readily established over a platinum catalyst[58]. In other cases the presence of the metal or oxide on the specimen surface is an effective catalyst, for example, iron, nickel and iron oxide. However, this catalytic activity may be poisoned by components of the gas mixture such as hydrogen sulphide[59].

The nature of a hot corrosion experiment requires that the gas mixture flows in a thermal gradient, so that care is required to ensure that thermal diffusion effects do not cause unmixing of the gas by differential diffusion of the components in the temperature gradient[60]. Generally the heavier components will tend to concentrate in the cooler regions of the furnace for high convection velocity and vice versa for low velocities. In addition, depletion effects can also cause changes in gas composition. In both cases it is advisable when starting a series of experiments in complex gas mixtures to determine whether the scaling behaviour depends on gas flow rate; when no such dependance is observed it can be inferred that both these effects have been reduced to negligable proportions.

b) Experimental Procedures

A paramount consideration is safety. Acceptable toxicity levels are given in the Registry of Toxic Effects of Chemical Substances[61], and the laboratory should be regularly checked to ensure that safe levels exist in the ambient atmosphere. This is best achieved by fitting the laboratory with gas detectors coupled to suitable warning devices preferably that isolate the gas supply system in the event of a serious leak. An adequate extraction system is, of course, also essential for the safe discharge of gases to the ambient atmosphere, or if this is undesirable the gas should be "scrubbed" of the noxious components (H_2S, for example). In addition, monitoring of inlet and outlet flow rates to the furnace gives an early indication of the development of leaks, and the use of double-walled furnaces which carry a flowing inert gas outer blanket can be used to remove the reaction gas safely should the inner tube fail. This latter procedure should be considered mandatory for combustible gases as it prevents direct contact with the hot furnace windings.

Procedures for starting an experiment must again take into account safety and also the establishment of the correct atmosphere. An important consideration is the use of inert gases to remove laboratory air from the system before introduction of the reaction gas; repeated evacuations with a rotary pump and backfilling is an alternative procedure. Specimens are either introduced into a cold furnace, the atmosphere established then heating to the test temperature, or alternatively, a gate-valve system can be used to allow samples to be directly introduced into the hot gas stream.

Monitoring the gas composition is, of course, desirable and gas chromatography and mass spectrometry are commonly used to determine the composition at ambient temperature, but it is not usually possible to measure gas compositions at the test temperature by these methods. Use of solid electrolyte zirconia probes to measure $p(O_2)$ of the gas mixture within the hot-zone is increasingly being used[62]. This method relies upon the establishment of the following electrolytic cell:

$$Pt/O_2 \ (air)/ZrO_2/Pt \ (O_2 \ unknown)$$

the emf generated is directly proportional to difference betweeen the reference (air) and unknown $p(O_2)$ values. These probes are available commercially or can be relatively easily made in the laboratory. There have been attempts to develop probes for the measurement of sulphur and carbon activities but these are still in their infancy[63].

5 ATTACK BY MOLTEN SALTS

The purpose of tests carried out to measure degradation induced by molten salts, or hot-salt corrosion, as it is commonly termed, is to determine the influence of the presence of a deposit of molten salt on the metal or alloy in an appropriate gas atmosphere. The rate of corrosion is controlled by the amount of salt present and by its chemistry, and any test procedure devised must attempt to achieve a realistic simulation of the service environment.

The case of gas turbine operation will be considered as an example. In this environment combustion of a sulphur bearing fuel in air containing sodium salt impurities (typically sea-water) results in the formation of a layer of Na_2SO_4 on the surface of the high pressure turbine blades. In designing a suitable test it should be borne in mind that the salt deposit is constantly being replenshed and that due to the action of centrifugal forces the deposit is also being removed to the blade tips and thus is usually quite thin. Although some molten salt may be trapped within pores of the oxide layer or a solid deposit which might also form on the blade. This is a complex situation which most researchers acknowledge cannot easily be realistically simulated in laboratory tests, so that various simplified test methods have been devised to allow some information about reactions kinetics and mechanisms to be obtained. Saunders and Nicholls[64] have reviewed the various test methods that have been used and classify these into crucible tests, furnace tests and burner rig tests which are briefly discussed below.

a)   Crucible tests.

This type of procedure involves immersion of specimens at an appropriate temperature in a molten salt based on sodium sulphate, but with additions of other alkali metal sulphates to form a low melting point eutectic mixture. The additions are necessary because test temperatures are usually lower than the melting point of sodium sulphate. In order to control the sulphidising potential (pS_2) in the environment to that characteristic of the combustion gas generated in service it is desirable to bubble air or oxygen containing SO_2/SO_3 mixture through the melt. The choice of the correct $p(S_2)$ ensures the stability of sulphates produced during the corrosion process and this is particularly important for low temperature hot corrosion (Type II) which depends on the formation of a $CoSO_4.Na_2SO_4$ eutectic which is stabilised by a relatively high $p(SO_3)$.

A major advantage of this type of test is that by using simple electro–chemical techniques it is possible to monitor the corrosion current and so obtain continuous measurement of the corrosion rate[65]. It is also possible to accelerate the corrosion process by anodically polarising the specimen. In a procedure designed by Shores[66] the specimens were "preconditioned" by polarising for about 2 hours at an anodic current of 3.17 mA cm^{-2} using in this case, a pure Na_2SO_4 melt at 906 °C. The time and temperature were somewhat arbitary and were chosen to be sufficient to initiate hot–corrosion. Anodic polarisation was then removed and the sample allowed to attain a stable rest potential, after which potentio–kinetic anodic polarisation was carried out to determine the corrosion current. This value of the corrosion current was taken by Shores to be a measure of the hot–corrosion resistance of the sample and the results obtained on a number of different alloys is shown in Figure 16 compared with results from burner rig tests, where it can be seen that a good correlation was obtained.

It must be emphasised that long–term exposure to molten salts in a crucible test may result in depletion of reactive species from the melt and this will modify corrosion results. Examination of the corrosion product morphology obtained from some of these tests, particularly those using salt mixtures whose composition departed from that expected in service is essential to determine if significant difference in corrosion process had occurred. Most of these tests only claimed to rank alloy performance with no attempt at obtaining quantitative kinetic data, and clearly do not attempt to simulate the service environment where salt replenishment by deposition occurs. Much thicker deposits are used in these tests than are found in service, so that it is not surprising that measured corrosion rates do not agree with service.

b) Furnace Tests

The Dean test[67] which involves both evaporation and deposition of sodium sulphate is probably the most appropriate furnace test in which control of $p(S_2)$ can also be effected, and overcomes some of the problems referred to in the previous section of unrealistic deposit compositions and thickness. The procedure involves having a reservoir of Na_2SO_4 at a high temperature and allowing Na_2SO_4 vapour to be carried into the gas stream to which the samples are then exposed at a lower temperature and condensation takes place to form a molten deposit on the sample. While this test is a better simulation of the service environment, the experimental arrangements are such that it is difficult to achieve the rates of deposition required to adequately reproduce the corrosion product morphology observed in service and rates of attack, particularly with more resistant materials are not comparable with engine experience.

Other variants of the furnace tests have been used and commonly samples are pre–coated with Na_2SO_4, usually to a weight of about 1 mg cm^{-2}, and placed in a furnace with a controlled atmosphere containing between 500 and 2000 ppm SO_2/SO_3 depending on the sulphur content expected in the combustion gas. The salt is usually replenished at intervals during the course of the test and typically daily renewal is a convenient compromise. An alternative procedure in which solid salt from a reservoir was continously fed onto the hot specimens – the salt shower test – was developed by Shaw[68]. In the case of these latter examples, it is relatively easy to alter the salt composition, often it is appropriate, for example to introduce NaCl as a component of the salt[68] or as Hossain and co–workers[69] to introduce HCl(g) into the atmosphere.

In general these tests are relatively easy to perform and do not require complex experimental equipment, but their disadvantage is that it is difficult to obtain continuous deposition that truely simulates rates found in operating engines.

c) Burner Rigs

The most generally accepted method of simulating operating conditions within a gas turbine is the use of burner rigs where combustion of fuel creates the test environment. A large variety of rigs exists, some are relatively simple using little more than a furnace with several independently controlled temperature zones[70,71], whereas others which operate at high velocity[25,26,72] and in a few cases, high pressure[73], are very complex and expensive to operate – in effect these rigs acquire many of the features of a gas turbine with large compressors for the supply of air and correspondingly high consumption of fuel. Usually contaminants, such as artifical seawater are added as an atomised stream to the secondary air introduced immediately after combustion of the fuel. In addition, vanadium compounds may also be added to the fuel to simulate use of poor quality residual fuels; Figure 5 shows a typical high velocity burner rig facility. The original concept of using high velocity rigs was to simulate closely the operating conditions within a gas turbine without running a trial engine.

Although there is general agreement that this type of test gives the best simulation of operating experience, and corrosion product morphologies obtained as a result of burner rig testing agree well with those found on blades taken from operating engines, considerable disagreement exists about the quantitative comparison of corrosion rates from different rigs. A major factor in the confusion is that in the past operators of rigs have attempted to compare results taken from rigs where the concentration of contaminants was controlled to that expected in the turbine. Since it has been established[64] that deposition rate is the primary factor controlling corrosion rates, only rigs operating at the same velocity as the turbine would be expected to produce comparable rates of attack. It has been pointed out[74,75] that contaminant flux rate (CFR) is a better parameter to control deposition and allows low velocity burner rigs to operate at comparable deposition rates to high velocity rigs and gas turbines. Currently, an international round–robin is taking place to establish the validity of this approach[76]. The principle of using equivalent CFR has been successfully used to simulate corrosion of exhaust valves in diesel engines[77].

6 ATTACK BY SOLID DEPOSITS

In cases where solid deposits form it is not only important to ensure that the amount and composition of the deposit correspond with that found in service, but, because of porosity effects in altering gas composition, the structure of the deposit must also be controlled.

Just et al[78], again in connection with materials selection for gas turbines, have developed a test in which a synthetic ash of composition similar to that found on the blades of gas turbines used in industrial processes covers the test sample and a gas atmosphere containing SO_2 is passed over the sample and ash. It is claimed that this procedure gives good ranking of performance, although rates of attack were generally higher than those encountered in service.

In recent years considerable interest has developed in fluidised bed combustion where deposit induced corrosion can occur. In this case the deposit consists of a mixture of $CaSO_4$ (derived from sulphur capture processes in the bed) and coal ash, where it is believed that the active species in the corrosion process is $CaSO_4$. A similar approach to that of Just et al has been adopted in this case with the sample being buried in various $CaSO_4$ containing mixtures ($CaSO_4+C$, $CaSO_4 + CaO$ etc) with or without control of the gas atmosphere. While at high temperatures these tests generally reproduced corrosion found in service they did not reproduce attack observed at temperatures below about 800 °C. This is believed to be due to inadequate simulation of the pore structure of the deposit. Slurries of $CaSO_4$ containing material have been used by Natesan[79] while Saunders and Spencer[80] attempted to overcome the problem by using powder compaction techniques. In these ash deposition conditions typical of those found within a fluidised bed, one cannot discount the role of partical impaction in modifying deposit morphologies and hence corrosion[81].

7 SYNERGISTIC CORROSION PROCESSES

a) Environment/Mechanical Load Interactions
Few components in service fail purely as a result of corrosion. Invariably, final failure results from some form of mechanical overload consequent upon loss of section due to corrosion. Hence corrosion processes play a major role in the initiation stages of many failure mechanisms, for example, fatigue cracks are often observed to initiate from corrosion pits, and may enhance the rates of mechanical failure. In this respect corrosion and mechanical

load interact synergistically.

Corrosion can reduce component sections, initiate cracks and modify the composition of the substrate through selective oxidation. This leads to increased creep rates, faster crack propogation rates and hence early component failure. Conversely, the transfer of strain from the substrate to any surface scale that is formed may result in scale fracture, spallation and hence loss in protection, further enhancing corrosion rates[82].

The enhanced failure rates observed when load and environment interact cannot be predicted from a knowledge of each component in isolation[83]. Hence studies of environmental creep, high temperature corrosion fatigue and high temperature stress corrosion cracking require the combination of mechanical load, corrosive environments, and possibly salt deposition in a single test procedure. Experimentally, these combined tests are undertaken by incorporating a high temperature corrosion autoclave on a conventional creep rig, tensile testing or fatigue machine. Testing methods proceed as for conventional high temperature mechanical tests, and this has been reviewed in other monographs in this series[84]. Remote monitoring of scale failure using acoustic emission can be incorporated, and it is useful to include a corrosion testpiece within the autoclave for direct comparison with the specimen exposed to combined stress and corrosion[47]. Both of these aspects are illustrated in Figure 12. For a more detailed discussion of facilities suitable for evaluating high temperature mechanical properties in aggressive environments reference should be made to recent conference proceedings on this subject[85,86,87].

b) High Temperature Wear Processes

The formation of compact oxide products within wear tracks is widely acknowledged to be beneficial[88,89,90,91], with the oxide glaze that is formed lowering the coefficient of friction and metal asperity contacts, hence reducing the observed wear rate. The morphology of these glazes is related to the high temperature strength of the alloy. When the alloy exhibits high strength a relatively thin glaze is formed, conversely for low strength alloys a comparatively thick oxide is observed[90]. In general, the thickness of the oxide that results is far greater than that expected under non-wearing, but oxidising conditions, and this is associated with the mechanical disruption of the alloy surface and oxide, resulting in accumulation of wear debris that forms the compact glaze.

Methods of assessing wear performance are reviewed elsewhere in this monograph[92]. Many of these machines have been modified to permit testing at temperatures up to 800 °C. One of the more popular rigs consists of a rotating disc or ring of material under study, which is in contact with either a pin or hemispherical rider. Other wear machines have used a reciprocating specimen plate in contact with a hemispherical button, (rubbing wear) or impact (hammer) wear varieties, with or without rotation. All of these machines can be configured to operate at temperature by incorporating the contacting materials within a furnace, or direct heating of the contacting surfaces using a gas flame or combustion products from a burner rig facility. Figure 17 illustrates the reciprocating wear test machine used by Stott and co-workers[88,89,90] to investigate high temperature wear.

c) Erosion/Corrosion Interactions

Solid particle impaction often contributes to the degradation processes observed in many high temperature systems. Particles may result from incomplete combustion or be entrained with the intake air, in gas turbine systems for example, or may result from the mode of plant operation as in coal fired fluidised beds and gasifiers.

As discussed earlier, the survival of high temperature alloys depends on the formation and maintenance of a protective oxide scale. Solid particle impaction can lead to scale fracture. Thus continual breakdown of the surface oxides by eroding particles allows contaminants access to the underlying metal resulting in enhanced degradation rates (erosion/corrosion).

Under conditions where the particle loading may be very high then erosion may be the dominant degradation mechanism, as observed on the leading edge of some turbines blades subjected to sand erosion[93,94]. Conversely, when particle loadings are low, little erosion is observed instead localised disruption of the protective oxide scales can lead to accelerated corrosion rates. For example, the inclusion of solid sea salt intermittently in a high velocity burner rig test at 700 °C resulted in local scale failure and enhanced corrosion with a classic Type II corrosion morphology. Local scale fracture is thought to reduce the incubation time for this form of attack by allowing contaminants localised access to the underlying alloy.

The foregoing has highlighted some of the complexities of erosion/corrosion interactions in high temperature systems. Mechanistic investigations and the assessment of materials performance therefore requires a wide range of testing techniques designed specifically to study different aspects of the erosion/corrosion processes at temperature.

Mechanistic studies aimed at evaluating the early stages of erosion/corrosion are best undertaken using a single particle impact approach[94,95,96]. Here, one or a small number of particles, entrained in a carrier gas, are fired at a preconditioned specimen surface at a desired test temperature, such that isolated impact damage sites can be evaluated. This technique permits particle–target interactions to be examined in detail under precise experimental conditions, hence, the influence of changes to the impact conditions on the material removal processes can be assessed. Such tests can be easily performed at temperature, in controlled environments and have been beneficial in understanding the role surface oxides play in high temperature erosion processes[96]. Single impact tests by definition cannot provide information on particle interactions or on steady state erosion rates.

To investigate particle interactions and the time dependence of erosion/corrosion processes discontinuous[94] (multiple impact studies) or continuous[97,98] erosion techniques are used. These methods involve a modification of the single impact gas gun approach to enable continuous or discontinuous entrainment of particles into the gas stream. Close control of experimental parameters is again possible and the influence of particle properties and environment on long–term erosion behaviour can be studied. Alternative methods based on rotating an inclined specimen through a particle stream[99] (the whirling arm rig) or centrifuging particles against a range of inclined specimens[100] have been used. The advantage of these later test methods is that they permit a wide range of environments to be tested using sealed autoclave systems.

A major disadvantage of these continuous erosion rigs is that they are unable to simulate the real environment. Contaminant levels and particle loadings are generally lower than those expected in service. Furthermore, because of the complex impact conditions encountered, it is often difficult to determine the exact erosion mechanism without some single impact studies.

More realistic simulations require laboratory rigs modelled on the expected plant conditions. Fluidised bed rigs[101,102,103] are used to simulated the in–bed erosion conditions expected in atmospheric and pressurised fluidised bed coal conversion plants. While high velocity burner rigs have been used to study high temperature erosion[104,105] or the influence of erosion on high temperature corrosion expected under turbine conditions[72]. These simulation rigs permit higher contaminant and/or particle fluxes and hence offer environmental conditions similar to those expected in service, provided that similar contaminant and particle fluxes are used. These rigs are good for ranking materials under the specific erosion–corrosion conditions used in the test, however less control over the exact impact dynamics and local environments means mechanistic studies are difficult. Even so, the use of combustion environments in many of these rigs provides important information on the interaction and long–term effects of erosion/deposition/corrosion processes.

Ultimately trials engines and pilot plant studies are the only true simulation of in service damage morphologies. However, these facilities are few in number and expensive to operate. A large degree of control over the exact environmental conditions is lost in operating the plant and hence it is difficult to correlate materials behaviour with particular operating parameters.

Clearly, no one technique can be used in isolation to investigate high temperature erosion/corrosion and it is only through a combination of approaches, that the mechanisms and long-term performance can be elucidated.

REFERENCES

1   P. Kofstad, "High Temperature Corrosion", Elsevier Applied Science, London, 1988.

2   O. Kubaschewski and B.E. Hopkins, "Oxidation of Metals and Alloys", Butterworths, London, 1967.

3   K. Hauffe, "Oxidation of Metals", Plenum, New York, 1966.

4   "High Temperature Corrosion", ed. R.A. Rapp, 1983, Houston Texas, National Association of Corrosion Engineers.

5   N. Birks and G.H. Meier, "Introduction to the High Temperature Oxidation of Metals" Edward Arnold, London 1983.

6   N.S. Bornstein and M.A. DeCrescente, Metall. Trans., 1973, 4, 261–278.

7   S.R.J. Saunders and S.J. Spencer, Mats. Sci. and Eng., 1987, 87, 227–235.

8   P.C. Rowlands and M.I. Manning, in "High Temperature Corrosion", (ed. R.A. Rapp), 300–309, 1983, Houston Tx, National Association of Corrosion Engineers.

9   J. Stringer, Corrosion Science, 1970, 10, 513–543.

10   P. Hancock and R.C. Hurst in "Advances in Corrision Science and Technology" Volume 4, (eds. M G Fontana and R W Staehle), 1–84, 1974, New York, Plenum Press.

11   J.V. Cathcart, "Stress Effects and the Oxidation of Metals", 1975, New York, Met. Soc. AIME.

12   Mater. Sci. Technol., 1988, 4, (5), (Special conf. issue – Oxide/Metal Interface and Adherence).

13   A.G. Evans and R.M. Cannon, in "Oxidation of Metals and Associated Mass Transport", (eds. M.A. Dayananda *et al*.),. 135–160, 1987, Warrendale PA, Met. Soc. of AIME.

14   P. Hancock, and   J.R. Nicholls, Mater. Sci. Technol., 1988, 4, 398–406.

15   H.E. Evans, Mater. Sci. Technol., 1988, 4, 415–420.

16   C. Wagner.   "Atom Movements", ASM Cleveland 1951, p.153–173.

17   W.E. Campell and U.B. Thomas, Trans. Electrochem. Soc., 1947, 91, 623–640.

18   D. Bruce and P. Hancock, J. Inst. Metals, 1969, 97, 140–148 and 148–155.

19   P. Hancock, Werkstaff u Korros, 1970, 21, 1002–1006.

20   R.C. Hurst, M.Davies and P. Hancock, Oxid. Met., 1975, 9, 307–355.

21   R.C. Lobb, Thermochemica Acta, 1984, 82, 191–200.

22   C.A. Barrett and C.E. Lowell, Oxid. Met., 1975, 9, 307–355.

23   C.A. Barrett, J.R. Johnston and W.A. Sanders, Oxid. Met., 1978, 12, 343–377.

24   H.E. Evans and R.C. Lobb, Corros. Sci., 1984, 24, 209–222.

25   C.H. Wells, P.S. Follansbee and R.R. Dils, in "Stress effects and the Oxidation of Metals", (ed. J.V. Cathcart), p220–244, 1975, New York, The Metallurgical Soc. of AIME.

26   H.J.C. Hersbach,   "High Temperature Facilities at NLR", Memorandum SL-88-005U, National – Zucht – en – Ruintevaartlaboratorium, Amsterdam, 1988.

27   A. Atkinson, R.I. Taylor and P.D. Goode, Oxid. Metals, (1979), 13, 519–543.

28   A.J. Rosenburg, J. Electrochem. Soc., 1960, 107, 795–9.

29   E.M. Fryt, Oxidation of Metals, 1978, 12, 139–56.

30   "Microstructural Characterisation", (ed. E Metcalfe), 1988, London, The Institute of Metals.

31   J.R. Nicholls and P. Hancock, "High Temperature Corrosion", (ed. R.A. Rapp), (1983), p.198–210 Houston, National Association of Corrosion Engineers.

32   J.R. Nicholls and P. Hancock in "Plant Corrosion Prediction of Materials Performance", (eds. J.E. Strutt and J.R. Nicholls), 257–273, Chichester, UK, Ellis Horwood, 1987.

33   H.E. Evans and R.C. Lobb, Proc. 9th Int. Cong. on Metallic Corrosion, Toronto, 1984, Vol 2, p46–53, National Research Council, Ottawa Canada.

34   J.R. Nicholls and P. Hancock in "The Role of Active Elements in the Oxidation of High–Temperature Metals and Alloys", (eds. E. Lang et al.), 1989, Elesvier – in press.

35   J.H. Stout and D.A. Shores, Mater. Sci. Eng., 1989, in press.

36   K.L. Luthra and C.L. Briant, Oxid. Met., 1986, 26, 397–416.

37   C. Roy and B. Burgess, Oxid. Met., 1970, 2, 235–261.

38   D. Delaunay, A.M. Huntz and P. Lacombe, Corros. Sci., 1980, 20, 1109–1117.

39   A.M. Huntz, Mater. Sci. Technol., 1988, 4, 1079–1088.

40   H.E. Evans and R.C. Lobb, Corros. Sci., 1984, 24, 209–222.

41   D.J Baxter, R.C. Hurst and R.T. Derricott, Werk. u Korros., 1984, 35. 266–272.

42   R. Rolls and M. Nematollahi, Oxid. Met., 1983, 20, 19–35.

43   P. Hancock in "Stress Effects and the Oxidation of Metals", (ed. J V Cathcart), p155–176, 1975, New York, Met. Soc. AIME.

44   C. Ilett, "Applications of Scanning Acoustic Microscopy in Materials Science", Oxford Univ, 1984. Thesis submitted for D. Phil.

45   A.S. Khanna, B.B. Jho, B. Raj, Oxid. Met., 1985, 23, 159–176.

46   W. Christl, A. Rahmel and M. Schütze, Mater. Sci. Eng., 1987, 87, 289–293.

47   M. Schutz, Oxid. Met., 1985, 24, 199–232.

48   J.B. Price and M.J. Bennett, Mater. Sci. Eng., 1989, in press.

49   F.K. Peters and H.J. Engell, cited in M. Schutze, Mater. Sci. Technol., 1988, 4, 407–414.

50   J. Stringer, Werkstoffe u Korros., 1972, 9, 747–754.

51   P.J. Burnett and D.S. Rickerby, Thin Solid Films, 1988, 157, 233–254.

52   A. Nicholson, cited in M R Wooton, "An Assessment of Methods Available to Measure Scale Metal Adhesion and the Mechanical Properties of Oxides", CEGB Report RB/D/N2116, November 1971, Berkeley Nuclear Labs, Berkeley, UK.

53   J.M. Hulley and R. Rolls, J. Iron and Steel Inst., 1970, 208, 1029–1030.

54   R. Rolls and F.V. Arnold, cited in P. Hancock and R.C. Hurst, "Advances in Corrosion Science and Technology", Volume 4, 1974, Plenum Press, p1–84.

55   R.H. Davies and T.I. Barry, "MTDATA Handbook", National Physical Laboratory, 1989.

56   K. Natesan, "High Temperature Corrosion" (ed. R A Rapp), 336–344, 1983, Houston Texas, National Association of Corrosion Engineers.

57   P. Kofstad, Mater. Sci. Eng., 1989, in press.

58   K.L. Luthra and W.L. Worrell in "Properties of High Temperature Alloys with Emphasis on Environmental Effects", (eds. Z.A. Foroulis and F.S. Pettit), 318–330, 1976, Princetown NJ, Electrochem Soc. Inc.

59   H.J. Grabke and I. Wolf, Mats. Sci. and Eng., 1987, 87, 23–33.

60   W. Jost, "Diffusion in Solids, Liquids, Gases", 492–501, 1952, New York, Academic Press Inc.

61   Registry of Toxic Effects of Chemical Substances, National Inst. for Occupational Safety, Cincinnati, Ohio.

62   K. Kiukkola and C. Wagner, J. Electrochem. Soc., 1957, 104, 379–387.

63   W.L. Worrell in "Metal–Slag–Gas Reactions and Processes", (eds. Z.A. Foroulis and W.W. Smeltzer), 822–833, 1975, Princetown NJ, Electrochemical Soc. Inc.

64   S.R.J. Saunders and J.R. Nicholls, Thin Solid Films, (1984), 119, 247–269.

65   A.J.B. Cutler and C.J. Grant, in "Deposition and Corrosion in Gas Turbines" eds. A.B. Hart and A.J.B. Cutler, Applied Science, London 1973, p.178–196.

66   D.A. Shores, Corrosion, 1975, 31, 434–440.

67   A.V. Dean, "Investigation into the resistance of various nickel and cobalt have alloys to sea salt corrosion at elevated temperatures", NGTE Report, January 1964.

68   S.W.K Shaw and M.T. Cunningham, "Initial High Temperature Corrosion Tests in a Modified Salt Shower Apparatus and Conventional Static and Cyclic Oxidation Tests", COST50, UK–9, INCO Progress Report, Birmingham, 1974.

69   M.K. Hossain, J.E. Rhoades–Brown and S.R.J. Saunders, "Behaviour of high temperature alloys in aggressive environments", Petten, 1979, (eds, I. Kirman et al.), Metal Society, London (1980) p.483–96.

70   P.A. Bergman, C.L. Sims and A.M. Beltran, "Materials Problems Associated with a Gas Turbine", 38– , ASTM, Spec. Pub. No 421, 1967, Philadelphia, Pa, Amer. Soc. for Testing and Materials.

71   S.R.J. Saunders, G.O. Lloyd and T.G. Dye, Proc. Conf. "High Temperature Alloys for Gas Turbines" (eds D Coutsouradis et al.), Liege, 25–27 Sept 1978, Applied Science Pub. London, 1978, 259–268.

72 P. Hancock in Proc. 4th Conf. on 'Gas Turbine Materials in a Marine Environment' Annapolis (1979) sponsored by US Naval Sea Systems Command Washington DC, USA, p.465-473

73 G.C. Booth, J.C. Galsworthy and A.F. Taylor, Proc. 4th Conf. "Gas Turbine Materials in a Marine Environment", Annapolis, 1979, sponsored by US Naval Sea Systems Command, Washington DC, p259-272.

74 P. Hancock, Corros. Sci., 1982, 22, 51-65.

75 S.R.J. Saunders, M.K. Hossain and J.M. Ferguson, Proc. Conf. "High Temperature Alloys for Gas Turbines 1982", Liege, 4-6 Oct 1982, eds. R Brunetaud et al, D Reidel, Publishing Co, Dordrecht, 1982, 177-206.

76 J.R. Nicholls and S.R.J. Saunders, High Temperature Technol., 1989, in press.

77 S.R.J. Saunders, S.J. Spencer and J.R. Nicholls, Proc. Conf. "Diesel Engine Combustion Chamber Materials for Heavy Fuel Operation", London, 26-27 Oct 1989, Inst. of Marine Engineers, in press.

78 C.H. Just, P. Huber and R. Bauer, Proc. 13th Int. Congr. on Combustion Engines, Council International des Machines à Combustion (CIMAC), Paris 1979, Paper GT 34.

79 K. Natesan, Corrosion, 1982, 38, 361-373.

80 S.R.J. Saunders and S.J. Spencer, Proc. Conf. "Performance of High Temperature Materials in Fluidised Bed Combustion Systems and Process Industries", Cincinnati, Oct 1987, ASM Int. 65-74.

81 D.J. Stephenson and J.R. Nicholls, 1989, Cranfield Inst. of Technology, unpublished work.

82 P. Hancock, G. Ward and B.S. Hockenhull, Met. Trans., 1974, 5, 1451-1458.

83 B.F. Dyson and S. Osgerby, Mats. Sci. and Tech., 1987, 3, 545-553.

84 "Physical and Elastic Charasterisation" ed. M McLean, 1989, London, The Institute of Metals.

85 I. Kirman *et al.*, "Behaviour of High Temperataure Alloys in agressive Environments", The Metals Society London, 1980.

86 D.J. Gooch and I.M. How, "Techniques for Multi–Axial Creep Testing", Elsevier Applied Science, 1986.

87 "Advances in Fracture Research (Fracture 81)" Volume 5, Cannes, France, March 1981, Pergamon Press, Oxford, 1982.

88 F.H. Stott and G.C. Wood., Tribology International, 1978, 11, 211–218.

89 J. Glascott, F.H. Stott and G.C. Wood, Oxid. Met., 1985, 24, (3/4), 99–114.

90 F.H. Stott, D.S. Lin and G.C. Wood, Corros. Sci., 1975, 13, 449–469.

91 T.F.J. Quinn, Wear, 1971, 18, 413–419.

92 T. Eyre (this monograph, Wear Methods).

93 R.L. Johnson, M.A. Swikert and E.E. Bisson, NACE, Tech Note 2758, 1952.

94 J.E. Restall and D.J. Stephenson, Mat. Sci. Eng., 1987, 88, 273–282.

95 D.J. Stephenson, J.R. Nicholls and P. Hancock, Corros. Sci., 1985, 25, 1181–1192.

96 D.J. Stephenson, J.R. Nicholls and P. Hancock, Corros. Sci., 1986, 26 (10), 757–767.

97 A. Levy, Wear, 1986, 111, 161–172.

98 L.K. Ives, J. Eng. Mater. Technol., 1977, 99, 126–135.

99 J.C. Galsworth, J.E. Restall and G.C. Booth, in Proc. Conf. "High Temperature Alloys for Gas Turbines 1982", (eds. R Brunetaud *et al.*), 1982, D Reidel Publishing Co. Dordrecht, 207–235.

100 S. Soderberg, S. Hogmark, U. Engham, H. Swahm, Tribology International, December 1981, 333–343.

101 M.J. Entwistle, I.M. Hutchings, J.A. Little, in Proc. 7th Int. Conf. on "Erosion by Liquid and Solid Impact", (eds. J E Field and J P Dear), Cavendish Laboratory, Cambridge, 1987, Paper 71, 1–10.

102 F.H. Stott, S.W. Green and G.C.Wood, Proc. Conf. Corrosion '89, New Orleans, April 1989, Nat. Ass. of Corrosion Engineers, paper 545, p1–13.

103 S. McAdam and J. Stringer, in Conf. Proc. "Wear of Materials", Denver, June 1989, ASME., 689–698.

104 R.H. Barkalow and F.S. Pettit, in Proc. Conf. on 'Corrosion/Erosion of Coal Conversion System Materials', Berkeley, California (1979) 139–173.

105 H.J. Kolkman, Proc. 6th Int. Conf. on Erosion by Liquid and Solid Impact, (Sept 1983) Cambridge UK, paper 45.

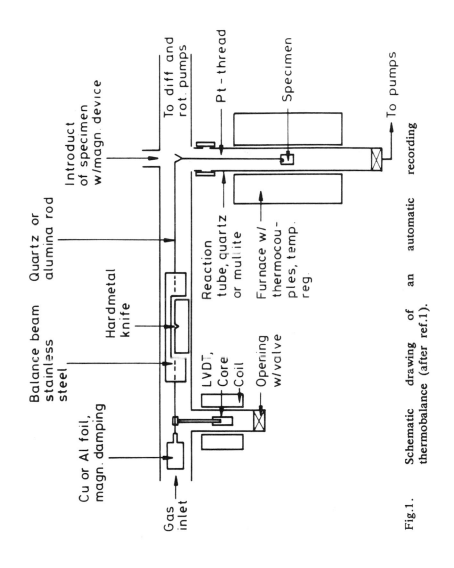

Fig.1.   Schematic   drawing   of   an   automatic   recording
thermobalance (after ref.1).

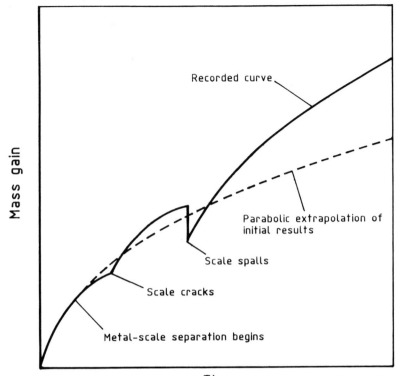

Fig.2.   Hypothetical mass gain versus time curve, showing exaggerated possible features revealed by continuous monitoring (after ref.5).

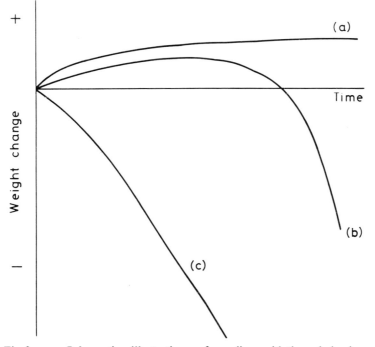

Fig.3. Schematic illustration of cyclic oxidation behaviour a) protective b) initially protective c) spalling.

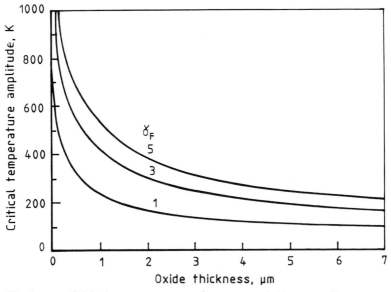

Fig.4. Critical temperature change to initiate spalling as a function of oxide thickness (after ref.33)

Fig.5.   Photograph of NLR cyclic burner rig facility (after
         ref.26).

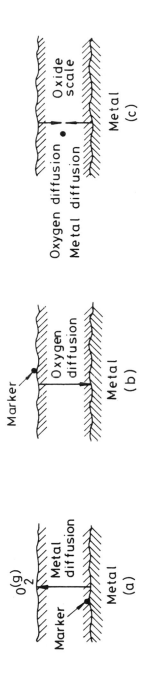

Fig.6. Ideal location of inert markers after lattice diffusion-controlled growth of oxide scales; a) metal diffusion predominant, b) oxygen diffusion predominant c) simultaneous metal and oxygen diffusion (after ref.1).

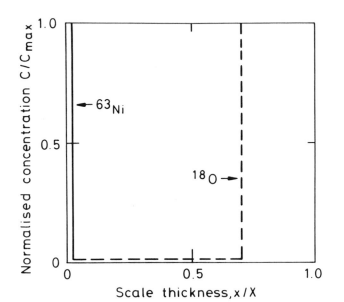

Fig.7a.   Distribution of metal tracer in a scale of MO growing by metal vacancy diffusion only. The metal tracer is deposited as a thin film on the metal surface prior to oxidation. The oxygen atoms in MO are considered to be immobile, and ^{18}O tracer introduced into the atmosphere during the last part of the oxidation is located in the outer part of the scale (after ref.27).

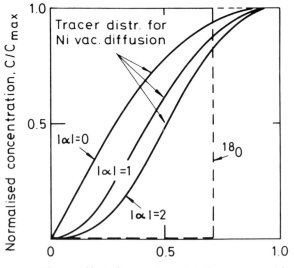

Fig.7b.   Distribution of M and ^{18}O tracers in MO scales growing by short–circuit (grain boundary) diffusion of the metal (after ref.27).

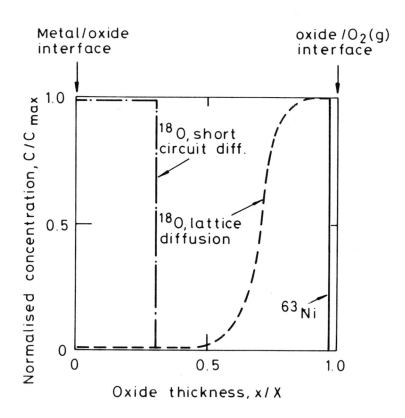

Fig.7c.   Distribution of M and ^{18}O tracers in MO scales growing by short–circuit and lattice diffusion (after ref.27).

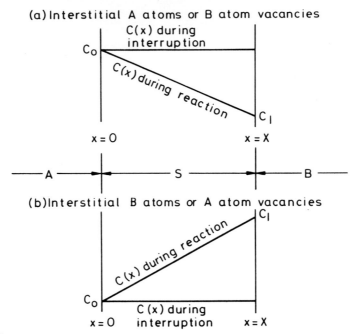

(a) Interstitial A atoms or B atom vacancies

C(x) during interruption

C_0

C(x) during reaction

C_1

x = 0   x = X

A —— S —— B

(b) Interstitial B atoms or A atom vacancies

C_1

C(x) during reaction

C_0

C(x) during interruption

x = 0   x = X

Fig.8.   Concentration of defects in a scale (S) interposed between A and B during the reaction aA + bB = S. In each case it is assumed that the supply of B can be interrupted while that of A is held constant (after ref.28).

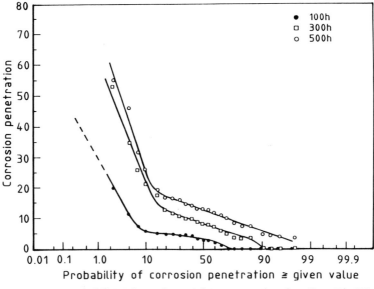

Fig.9.   Probability plots of corrosive penetration in alloy IN 939 after burner rig testing at 700 °C (after ref.32).

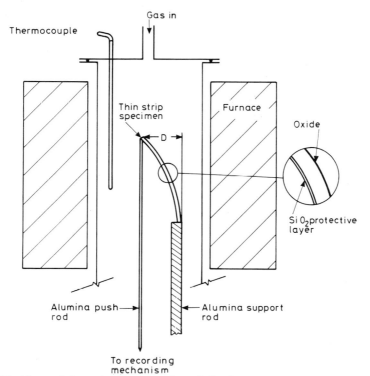

Fig.10.   Schematic diagram of a deflection apparatus to measure strain in growing scales (after ref.37).

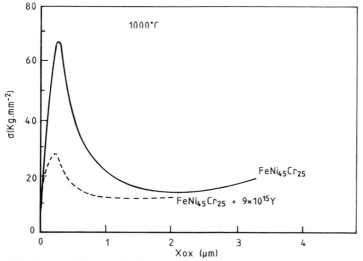

Fig.11.   Values of the compressive stresses developed in the oxide scale of Fe.45Ni.25Cr and Fe.45Ni.25Cr.Y alloys versus oxide thickness, X_{ox} (after ref.39).

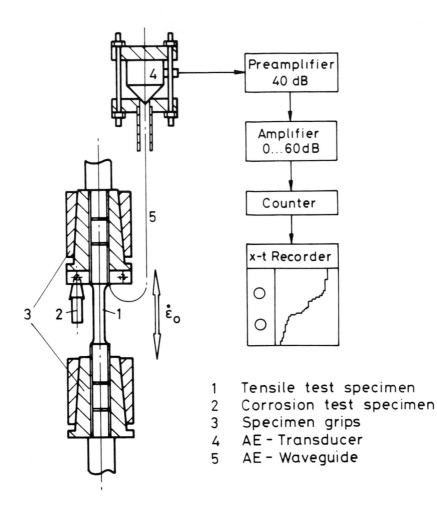

Preamplifier
40 dB

Amplifier
0...60dB

Counter

x-t Recorder

$\dot{\varepsilon}_0$

1 Tensile test specimen
2 Corrosion test specimen
3 Specimen grips
4 AE - Transducer
5 AE - Waveguide

Fig.12.   Schematic diagram of an acoustic emission apparatus
coupled to a tensile test equipment (after ref.47).

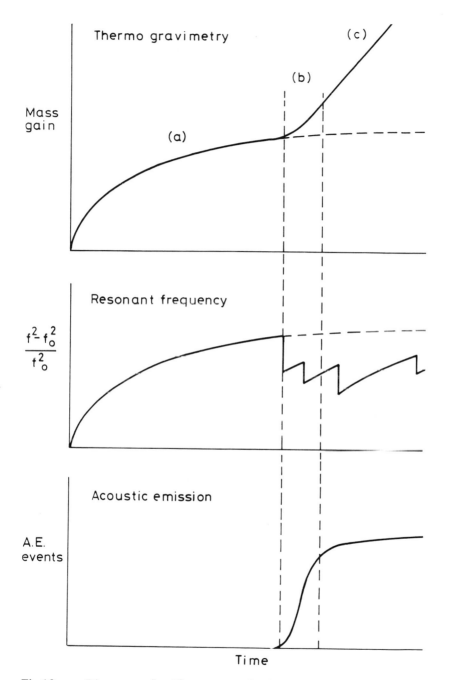

Fig.13.   Diagrammatic illustration of changes in specimen mass, acoustic emission and resonance frequency as a function of time.

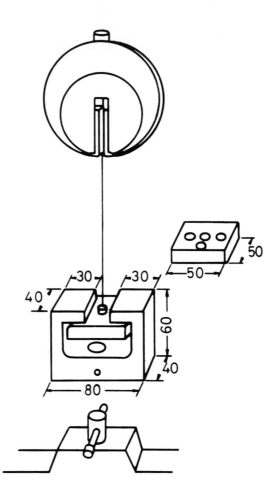

Fig.14.　Schematic diagram of a pull adhesion test arrangement (after ref.49).

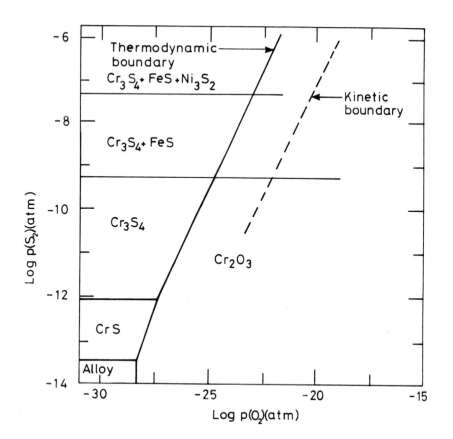

Fig. 15. Phase stability diagram for the Cr–O–S system on Incoloy 800H at 1023K showing thermodynamic and kinetic boundaries (after ref.56).

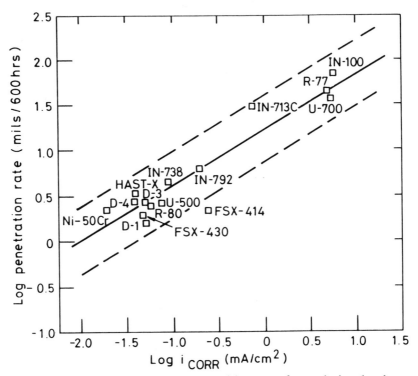

Fig.16. Correlation of $\log_{\text{icorr}}$ with rate of attack in the burner rig by linear regression analysis. The dashed lines are the 95% confidence limits (after ref.66).

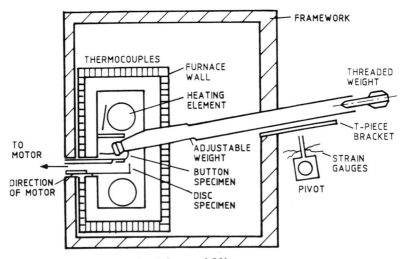

Fig.17. Hot wear rig (after ref.90).

144

4:Oxide characterisation for life prediction
E METCALFE

Dr. Metcalfe is in the Materials and Welding Section of
the Engineering, Science and Technology Department of
National Power Company which is based at Leatherhead,
Surrey, UK.

1. **INTRODUCTION**

During routine service inspections of high temperature
components cracks are sometimes found. The main
questions to be answered are what caused the crack, how
long has it been present, and is it still propagating?
Once these questions are answered decisions can be made
about whether to replace the component or repair it,
and about the inspection strategy for other components.
For many years fractography has been an important
technique for service failure investigations and
prediction of the life of components. Examination of
the fracture surfaces has provided information about
the site of the initiation of a defect, its mode of
propogation and in certain circumstances, the influence
of operating parameters, such as cyclic loading, on the
growth of the defect. For power plant components,
however, the high temperature environments usually lead
to the formation of a thick metal oxide scale on all
exposed surfaces. The oxide scale may mask all
fractographic evidence partly because of the
consumption of the underlying metal during oxide
growth.
        However there is still a lot of information that
can be obtained from the examination of the oxide on

the surface of a component during a service failure investigation. Examination of the oxide, coupled with a detailed knowledge of a relationship between oxide properties and the service environment can provide valuable information on the thermal history of the component. This has come to be known as 'oxide dating' and has been used to provide vital information on the age of cracks or the operating temperature of a component. Background data on the variation of oxidation kinetics, oxide morphology, phase composition, solute concentration and lattice parameter as a function of time and temperature of exposure, surface condition and defect concentration have been determined over a number of years for low alloy steels. These data can provide methods by which the age and environmental/thermal history of a failed high temperature component can be accurately determined in the absence of direct metallographic or fractographic evidence. The number of service applications of this technique has increased in recent years, and the technique as utilized by the CEGB has been reviewed by Pinder[1] and Metcalfe et al[2].

The principles of oxide characterisation will be described followed by a detailed description of the oxide dating technique. Examples of case studies of typical failure investigations will be given followed by an assessment of the errors in the technique.

2. **PRINCIPLES OF OXIDE CHARACTERISATION**

The growth of oxide scales on steels at high temperatures takes place by diffusion of iron out through the scale to the scale/oxidant interface and the diffusion of oxygen in through the scale to the scale/metal interface. Because it is a diffusion controlled process a plot of a measure of the oxidation rate, such as oxide thickness or weight gain, as a function of time generates a parabolic curve (Figure 1) obeying the relationship

$$x^2 = K_p t \qquad (1)$$

where x = oxide thickness, K_p = parabolic rate constant and t = exposure time.

Since diffusion is a thermally activated process changes in temperature will result in changes to the value of K_p and this follows an Arrhenius relationship

$$K_p = K_o \exp -\frac{\Delta G}{RT} \qquad (2)$$

where K_o = Arrhenius constant, ΔG = activation energy, R = gas constant and T = temperature. The temperature dependence of the rate of oxidation can be represented by an Arrhenius plot of log K_p versus the reciprocal of temperature. Thus if the oxide thickness and temperature are known time can be calculated and vice versa. A consequence of the time and temperature dependence of oxidation is that any given oxide thickness may be attributable to any combination of exposure time and temperature which satisfies the relevant kinetic data. It is often useful to find some way of putting limits on one of these parameters and the easiest way is to examine how temperature affects the morpholgy of the oxide.

2.1 Oxide Morphology

Steels used for high temperature components range from low alloy steels such as mild steel and $\frac{1}{2}$CrMoV steel, through the higher chromium ferritics such as 9%Cr 1%Mo steel and 12%Cr 1%Mo steel to the austenitic steels such as Type 316 which contain 18%Cr and 11%Ni. Much of the oxidation data for such steels has been gathered in either air or steam and the basic mechanism of oxide scale growth is the same for both types of steel.

During oxidation of ferritic steels a two-layer oxide usually forms. This is shown schematically in Figure 2. The interface between the two layers is the original metal surface. The outer layer consists of pure iron oxides, the phases present being a function of temperature and oxygen partial pressure. The most common oxide formed is the spinel phase Fe_3O_4, and it grows in the form of long columnar crystals. The scale grows by the diffusion of iron in the form of Fe^{2+} or Fe^{3+} through the oxide to the scale/oxidant interface. The inner layer consists of a spinel structure M_3O_4 where M comprises both Fe and the alloying elements. The inner layer occupies the same volume as the steel from which it is formed and the removal of iron to form the outer layer means that the alloying elements become more concentrated in the inner layer spinel compared to the underlying metal. The grain structure of the inner layer is very fine and is often difficult to measure (it can usually only be determined using electron microscopy). Figure 3 is an optical micrograph of a polished section through a scale formed on 9%Cr 1%Mo steel in steam, and shows the duplex nature of the scale. Figure 4 is a scanning electron micrograph of a fracture through the same scale and shows the outer layer/ inner layer interface in detail. The columnar structure of the outer layer is clearly seen and the fine structure of the inner layer should be noted.

This basic morphology applies to all scales but considerable variations arise depending on the composition of the steel, the temperature, and the oxygen partial pressure. Some of these variations can assist in determining the temperature of formation and Pinder[1] has carried out a detailed assessment for low alloy steels. On pure iron different oxide phases of varying proportions may form the final scale thickness, depending on the exposure temperature. For instance, the scale formed below 570°C consists of a layer of magnetite (Fe_3O_4) overlaid with a layer of haematite (Fe_2O_3). This is a higher form of iron oxide and forms as a layer on top of the magnetite when the oxygen partial pressure of the oxidant phase is sufficiently high. Above 570°C wustite (FeO), comprising up to 90% of the total scale thickness is formed between the metal substrate and the magnetite layer[3]. The relative thicknesses of individual phase layers may therefore be of value in determining exposure temperature. Unfortunately for pure iron and mild steels the relative ratios of individual phase thicknesses are independent of temperature above and below 570°C. The presence or absence of wustite can then only be used to indicate whether or not a component has operated above 570°C.

For low chromium alloy steels, however, the relative thicknesses of magnetite and haematite are temperature sensitive (Figure 5). Above 650°C wustite becomes stable, forming a fourth layer interposed between the inner-layer spinel and the outward growing magnetite and haematite layers. Thus measurement of the individual phase thicknesses and observations of the presence of the inner-layer spinel and wustite may be used to estimate the exposure temperature of a component from the oxide morphology. Above 650°C, however, the relative thicknesses of the four oxide layers within the scale are again insensitive to formation temperature.

2.2 Oxide Lattice Parameter

The lattice parameter of wustite varies as a function of iron concentration, and this, combined with the variation of stoichiometry with temperature, indicates that measurements of oxide lattice parameter may be used to estimate the oxide formation temperature. Measurements of the lattice parameters of wustite grown on mild steel and low chromium alloy steels between 650°C and 850°C show a continuous decline of lattice parameter with increasing temperature of formation[1] (Figure 6). The values obtained tend to fall on or around the calculated average lattice parameter of wustite in equilibrium with iron and magnetite assuming a linear concentration gradient of iron between the two phase boundaries.

For magnetite grown on mild steel the lattice parameter is insensitive to formation temperature between 500 and 850°C (Figure 7) because of its very narrow limits of stoichiometry within this temperature range. For low chromium alloy steels, however, at temperatures below 650°C, the segregation of the chromium into the inner layer spinel results in a decline of the magnetite lattice parameter to a value similar to that observed on mild steel. The lattice parameter of the magnetite is then insensitive to temperature[4].

In summary, therefore, lattice parameter measurements are only of use at the high temperatures where wustite might form on low alloy steels. For the majority of cases where only magnetite and haematite form such measurements will not contribute to an estimation of temperature of formation.

2.3 Non-Parabolic Kinetics

The determination of either exposure temperature or

exposure time depends upon confidence of the knowledge of the oxidation kinetics. If the kinetics for the actual component under investigation are not the same as those which have been determined for the oxidation data base then large errors can result. Operational factors such as thermal cycling or repeated mechanical stress loading can lead to accelerated oxidation rates. For example, thermal cycling of a 2¼Cr 1Mo steel has been demonstrated to give enhanced corrosion rates[5]. When this occurs, however, careful metallographic examination usually reveals the existence of laminated scale growth which can have kinetics rangeing from parabolic to linear. Such laminated scales consist of pairs of layers of magnetite and iron-chromium spinel. The magnetite layer at the steam/scale interface is considerably thicker than any of the other layers and is paired with a spinel layer which is half its thickness. The component layers of the remaining pairs are of roughly equal thickness, and although the thickness of the pairs is found to vary from <1 μm to 50 μm, no consistent trend in this variation across the scale is seen. Metallographic examination of the scales indicates that growth of the oxide laminations takes place at the metal interface and immediately adjacent to it underneath the oxide layers already formed. Thus the mechanism of growth of the pairs of layers is probably similar to that of the duplex oxide scales formed on low chromium steels in several oxidation environments such as high temperature water[6] and steam[7]. Growth of the magnetite layer takes place via the outward diffusion of iron through the oxide scale with the formation of new oxide at the outer magnetite surface. In laminated scales this growth will be at the interface between the bottom of the spinel layer of the previous pair of laminations and the top of the magnetite layer of the currently growing pair. The spinel layer in a pair of laminations grows by the inward diffusion of oxygen to the metal/scale interface and effectively fills the volume left behind by the iron which has formed the outer layer. The extra thickness of the outermost magnetite layer over the spinel layer with which it is paired probably arises through the continued diffusion of some of the iron to the steam/scale interface during growth of the laminations. The morphology of such an oxide layer is quite distinctive (Figure 8) and careful metallography can identify those cases where abnormal oxide growth has occurred and the application of oxide dating is inappropriate. Figure 9 shows the effect of different lamination widths on the rate kinetics for oxidation with a very finely laminated oxide tending towards linear growth kinetics.

3. OXIDE DATING

The kinetics of oxide growth were briefly introduced in Section 2. In order to appreciate the strengths and limitations of the oxide dating technique a fuller description of the growth kinetics is required. Duplex oxide growth on steels is generally found to obey a kinetic growth equation of the form

$$x = Kt^m \qquad (3)$$

where x = oxide thickness, K = rate constant, t = time and m is the exponent which is often 0.5 (parabolic growth law) but can vary from 0.3 to 0.7 depending on the alloy and the exposure conditions. As shown earlier K is generally related to the absolute temperature by an Arrhenius relationship

$$K = K_o \exp - \frac{\Delta G}{RT} \qquad (4)$$

where K_o = Arrhenius constant, ΔG = activation energy, R = gas constant and T = temperature. If the constants in these relationships are known from laboratory measurements on the same steel, the exposure time t can be calculated if the oxide thickness is measured from plant components and T is known or an average T can be calculated if t is known. If both T and t are unknown at least constraints can be placed upon them by measurements of the oxide thickness and detailed assessment of the oxide morphology may assist in providing further constraints.

In using equation (3) it should be realised that K and m are correlated[8] and a simple way of expressing this is by an equation of the form

$$m = \alpha + \beta \ln K \qquad (5)$$

Therefore we may substitute equations (4) and (5) into equation (3) and obtain

$$x = K_o \exp - \frac{\Delta G}{RT} \, t^{(\alpha + \beta \ln K)} \qquad (6)$$

Letting $G = \Delta G/R$

$$x = K_o \, e^{-G/T} t^{\left(\alpha - \frac{\beta G}{T} + \beta \ln K_o\right)} \qquad (7)$$

and

$$t = \left(\frac{x}{K_o \, e^{-G/T}}\right)^{\left[\frac{1}{\alpha - \beta G/T + \beta \ln K_o}\right]} \qquad (8)$$

This is the basic equation for calculating the time required to grow an oxide of thickness x at a temperature T. It is not as formidable as it first looks. K and m are not always correlated or there is not sufficient data to test the correlation. This leads to simplification of the above equation.

3.1 Oxidation at More Than One Temperature

When oxidation occurs at more than one temperature the method of calculation is somewhat different. An example is shown in Figure 10. A metal component has been exposed at a temperature T_1 for a time t_1 and then at a lower temperature T_2 for a time t_2. Sometime during the total exposure period $t_1 + t_2$ a crack has formed and its surface oxide has a thickness x_c. It is required to find the lifetime of the crack t_c. Both of the operating temperatures and both of the times at these temperatures have been recorded during the lifetime of the component with reasonable accuracy.

If the crack had formed during exposure at the lower temperature T_2 then from equation (3);

$$t_c = \left(\frac{x_c}{K_2}\right)^{1/m_2} \tag{9}$$

However, if t calculated in equation (9) is greater than t_2, the crack formed in the earlier period t_1. Therefore we have to obtain the 'equivalent time' for exposure at T_2 to form an oxide of thickness x_c. This is given by;

$$t_{2eq} = \left(\frac{x_c}{K_2}\right)^{1/m_2} \tag{10}$$

Oxidation at temperature T_2 for the time $t_{2eq} - t_2$ would give an oxide thickness x' at the start of the oxidation period at T_2 which is given by;

$$x' = K_2 \left(t_{2eq} - t_2\right)^{m_2} \tag{11}$$

This must be equal to the oxide thickness at the end of the oxidation period at T_1 (assuming that there are no significant re-equilibrium kinetics). Therefore the time Δt over which oxidation occurred at T_1 is given by;

$$\Delta t = \left(\frac{x'}{K_1}\right)^{1/m_1} \tag{12}$$

Substituting for x' and t_{2eq} in equation (12) gives;

$$\Delta t = \left(\frac{K_2}{K_1}\right)^{1/m_1} \left[\left(\frac{x_c}{K_2}\right)^{1/m_2} - t_2\right]^{m_2/m_1} \tag{13}$$

If $m_1 = m_2 = m$ this may be simplified to give;

$$\Delta t = \left(\frac{x_c}{K_1}\right)^{1/m} - t_2 \left(\frac{K_2}{K_1}\right)^{1/m} \qquad (14)$$

The total defect age will therefore be $t_2 + \Delta t$.

This procedure may be extended to include a range of temperatures and times. Therefore if the oxide thickness on a crack face is measured and there is sufficient laboratory data to determine the rate constants at the temperatures of interest the total age of the crack can be determined.

3.2 Oxide Blocking of Defects

One of the common problems encountered during oxide characterisation of service failures is the complete oxide blocking of the defect under investigation. In these circumstances a radical modification of the oxidation kinetics and oxide morphology will occur. To attempt to use the oxide in a blocked crack for oxide dating purposes can lead to great uncertainties. However the morphological changes which occur in the oxide allow a crack which has been fully blocked with oxide to be easily identified, and disregarded for oxide dating purposes.

When the outer layers of scales growing on opposite sides of a crack impinge, the path through the outer layer from the centre of the crack to the outer layer/oxidant interface will increase rapidly to a length which is much greater than the thickness of the inner layer (see Figure 11). Iron will no longer be deposited in the crevice, since this is a site of relatively low oxygen partial pressure compared with the scale/gas interface. This increase in the path length of magnetite through which both iron and oxidant must diffuse will reduce the oxidation rate in the crevice relative to that of the free surface. With increasing time the rate of oxidation will become negligible.

If this process occurs in an oxidation atmosphere of a sufficiently high oxygen partial pressure then down the centre of the blocked crack there will initially be a layer of haematite. The continued flux

154

of iron ions from the metal substrate, down the established activity gradients within the scale will cause the reduction of the haematite layer by the normal magnetite growth mechanism (Figure 12). On achieving full reduction of the haematite it is possible for the magnetite itself to be reduced by the carbon in the substrate alloy leading to a band of pure iron being formed in the centre of the crack. The loss of carbon from the alloy produces a decarburised zone around the crack which is easily seen when etched.

There is therefore usually sufficient evidence obtained by metallography to determine if a crack is blocked and when oxide dating cannot be applied. Attempts have been made to use the reduction kinetics for oxide scales to obtain estimates of the age of the crack but this is not yet an accepted technique because of the uncertainties of the transition between oxidation and reduction.

3.3 **Application of the Technique**

The way in which the oxide dating technique is used to estimate the exposure time of a component depends upon the quantity and the quality of the data which are available on the oxidation characteristics of the same steel under laboratory conditions. Three possible analytical routes are considered below for different quantities of available data.

CASE 1 A single set of data (weight gain versus time for one sample) or multiple sets of data (weight gain versus time for a number of samples) exist at the single temperature of interest.

This is a straightforward case where both K and m may be easily obtained and only an extrapolation over time is required.

CASE 2 Single sets of data exist at different temperatures allowing extrapolation or interpolation to a particular temperature of interest.

In this case both K and m may be obtained by a plot of ln x versus ln t and then ln K may be plotted against 1/T to obtain the temperature dependence of K. Note that there are not sufficient data to determine the correlation between m and K.

CASE 3 Multiple sets of data exist at different temperatures allowing extrapolation or interpolation to a particular temperature of interest.

In this case the correlation between m and K can be tested. If the slopes of the curves for ln K versus m are not significantly different and the mean m values do not change significantly with temperature, then one can return to Case 2. If not then a more complicated analysis must be performed as it implies that m varies with temperature. The treatment of the data can be simplified by making an assumption about the form of the temperature dependence. As ln K is a function of $1/T$ and of m these may be linearly combined into the simple relationship;

$$\ln K = A + Bm + \frac{C}{T} \tag{15}$$

Substituting for K in equation (3) gives;

$$x = \exp\left(A + Bm + \frac{C}{T} \right) \cdot t^m \tag{16}$$

or

$$\ln x = A + Bm + \frac{C}{T} + m \ln t \tag{17}$$

This must be solved using a non-linear regression technique.

4. **EXAMPLES OF ANALYSES**

Oxide dating has played a valuable role in many failure investigations and has become one of the main ways of determining the age of defects in high temperature plant. Sometimes during regular plant inspections a defect is found, often associated with a weld, using non-destructive testing techniques which are superior to those employed when the plant was built. For

156

example the defect may take the form of a small crack emanating from the unfused land of a butt weld in the structure. If it can be shown that the crack is of the same age as the weld and was caused during post-weld heat treatment, and has not propagated during service, this increases the confidence in the structure.

Often oxide dating has a role to play in more extensive investigations. Ten years ago the studs in a governor valve in an Intermediate Pressure (IP) turbine in a CEGB power station failed, resulting in the explosive detachment of the governor valve which was ejected through the roof of the turbine hall. Figure 13 shows a diagram of the steam chest of the IP turbine showing the location of the governor valve (the emergency stop valve is 560 mm in diameter and the governor valve is 508 mm in diameter). Figure 14 is a photograph of the steam chest after the incident. No people were injured during the incident. The investigation into the cause of the failure centred on an assessment of the 24 retaining studs which were manufactured from Durehete 1055, a high strength ferritic CrMoV bolting alloy containing additions of Ti and B, which is widely used in power plant at temperatures of up to 570°C. Such studs have a hollow centre to allow heaters to be inserted to cause stud expansion during retightening.

One of the main causes of the failure was inadequate lagging leading to a temperature differential along each of the studs which caused bending of the studs. This increased the strain in the region of the first engaged thread of the stud. This, combined with the loss of creep strength in some of the studs, led to the rapid failure of some of the studs, thus transferring more load onto the others. These failed by tensile rupture due to overload.

Oxide dating had two roles to play in the investigation; first to determine if there was a temperature gradient along the studs, and second to estimate the time of fracture of the failed studs. Oxide thickness measurements were taken along the centre hole in each of the studs. The operating time of the unit was known (54,000 h) and so a temperature profile along the central hole of each of the studs could be obtained. Figure 15 shows such a profile indicating that a temperature gradient did exist. This was supported by hardness measurements (note that these only indicated a gradient and not the actual temperatures) and by later actual measurements on plant using specially instrumented studs. During the oxide dating assessment some of the studs gave anomalously low mean temperature values when the temperature was calculated using the mean unit operating time. When the calculation was done using the last outage time (15,000 h, when some studs were replaced) the same stud operating temperatures were obtained. Thus the studs which were replaced at the last outage could be identified.

Oxide thickness measurements were also made across the fracture surface of selected studs, and the values were converted to crack age estimates. The results showed that within the accuracy of the technique (± 1000 h) the crack initiated just prior to or during the last period of operation and generally took several thousand hours to propagate from the outer edge to the bore of the hollow stud, indicating a process of slow crack propagation.

Oxide dating has also been used for more detailed assessments of crack growth[1]. For a surface emergent defect growing slowly with time, the final oxide thickness measured on the defect surfaces will decrease from the crack mouth to the crack tip provided that the crack is not fully oxide blocked. An example of this variation in oxide thickness with crack depth is shown in Figure 16. A surface emergent thumbnail-shaped defect was found to be growing into a plain carbon steel pipe from the outer surface. The pipe outer surface had been exposed to air at 360°C for 59,000 h. The defect surfaces had also been exposed to air for an unknown period. Calculation of the age of the defect from oxide thickness measurements taken every 1 mm down the crack length produced a curve of crack length as a function of operating hours as shown in Figure 17. Progressive growth of the defect with time and acceleration of the crack propagation rate with increasing crack depth are both indicated by the shape of the curve in Figure 17. Further analysis may be carried out. Several methods are available for relating parameters such as reference stress, crack tip stress intensity, K etc to different crack growth mechanisms. By constructing tangents to the crack length/time curve (Figure 17) a relationship between crack propagation rate and crack length can be obtained as shown in Figure 18. Such information can be used to assess the principle method of crack propagation.

5. ERRORS IN THE TECHNIQUE

The number of service applications of the technique has increased in recent years, and improved oxidation data are being collected which will improve the accuracy of the technique. However general acceptance of the oxide dating technique has been slow because of three reservations about its use:

1. When examining oxide thicknesses along cracks in components, crack surface topography and restrictions in oxidant access may change the oxidation rate relative to that of a free surface.

2. Operational factors such as thermal cycling or repeated mechanical stress loading can lead to accelerated oxidation rates.

3. It was not fully known how errors in each of the variables involved in the calculation affect the error on the final result, particularly when extrapolation or interpolation of short term laboratory data is necessary.

The first reservation has been considered by Pinder[1] who monitored the oxidation behaviour of fatigue fracture, brittle cleavage fracture, and emery ground surfaces of mild steel. During the early stages of exposure (50 to 100 h) fracture surfaces exhibited enhanced oxidation weight gain kinetics attributable to the surface roughness factor and cold work effects. Long exposure (>100 h) produced a levelling in the oxidation front and a reduction of cold work effects resulting in a decline of the instantaneous rate constant to that observed on flat surfaces. On fracture surfaces the oxide thicknesses, after several hundred hours of exposure, were equal (within experimental error) to those on emery ground surfaces. The oxide morphology and crack growth kinetics prior to and following oxide blockage of cracks in mild steel and CrMoV in air have also been studied[1]. Prior to oxide blockage the oxide morphology and growth kinetics were consistent with those observed on free unrestricted surfaces. No effects of any possible gas composition change with crack width or depth were observed. It was concluded that with regard to oxide characterization for service failure diagnosis, the effects of these factors are small.

The second reservation can be important because thermal cycling has been demonstrated to give enhanced corrosion rates[5]. When this occurs however, careful metallographic examination reveals the existence of laminated scale growth which can have kinetics rangeing from parabolic to linear (see Section 2.3). Thus careful metallography should identify those cases where abnormal oxide growth has occurred and the application of oxide dating is inappropriate.

The third reservation is one which is often raised and so it is appropriate to deal with it in some detail. It is important to be able to quote the correct errors for the technique and to appreciate the possible sources of error and the ways in which the oxidation data may be collected and analysed to minimise such errors.

5.1 Error Propagation Analysis

Analysis of the errors which can arise during the application of the oxide dating technique using available data can be treated using error propagation analysis assuming Gaussian statistics. If $x = f(u,v,w)$ then the standard error in x is given by[9];

$$\sigma_x^2 = \sigma_u^2 \left(\frac{\delta x}{\delta u}\right)^2 + \sigma_v^2 \left(\frac{\delta x}{\delta v}\right)^2 + \sigma_w^2 \left(\frac{\delta x}{\delta w}\right)^2 \qquad (18)$$

where σ_i is the standard error in the value of the variable i. Note that u, v and w must be uncorrelated.

Using this expression the error in t can be determined for the three cases defined in the section 3.2. The actual derivations of the equations below are given in Metcalfe, Taylor and Broomfield[2].

CASE 1. In this case for a single set of data at one temperature, equation (3) is used and σ_t is given by;

$$\frac{\sigma_t^2}{t^2} = \frac{\sigma_x^2}{x^2} \cdot \frac{1}{m^2} + \frac{\sigma_m^2}{m^2} (\ln t)^2 + \frac{\sigma_K^2}{K^2} \cdot \frac{1}{m^2} \qquad (19)$$

CASE 2. When single sets of data exist at different temperatures with extrapolation or interpolation being required to a particular temperature of interest the error in t can be determined by combining equations (3) and (4) to give;

$$t = \left[\frac{x}{K_o \exp(-\frac{G}{T})} \right]^{1/m} \qquad (20)$$

However to obtain a value for K_o a long extrapolation to $1/T = 0$ would be required. To avoid this we redefine K as;

$$K = K_1 \exp - G(\frac{1}{T} - \frac{1}{\bar{T}}) \qquad (21)$$

where $\bar{T}$ is the arithmetic mean of the temperature range of interest. The error in t is then given by;

$$\frac{\sigma_t^2}{t^2} = \frac{\sigma_x^2}{x^2} \cdot \frac{1}{m^2} + \frac{\sigma_m^2}{m^2} (\ln t)^2 + \frac{\sigma_{K_1}^2}{K_1^2} \cdot \frac{1}{m^2} + \frac{\sigma_T^2}{T^2} \cdot \frac{G^2}{T^2 m^2} + \frac{\sigma_G^2}{m^2} \left[\frac{1}{T} - \frac{1}{\bar{T}} \right]^2$$

$$\dots (22)$$

CASE 3. When multiple sets of data exist at different temperatures with extrapolation or interpolation being required to a particular temperature of interest the error in t is obtained by use of equation (17);

$$t = \left[\frac{x}{\exp(A + Bm + \frac{C}{T})} \right]^{1/m} \qquad (23)$$

The result gives;

$$\frac{\sigma_t^2}{t^2} = \frac{\sigma_x^2}{x^2}\cdot\frac{1}{m^2} + \frac{m}{m^2}(\ln t)^2 + B^2 + \frac{\sigma_A^2}{m^2} + \sigma_B^2 + \frac{\sigma_C^2}{T^2}\cdot\frac{1}{m^2} + \frac{\sigma_T^2}{T^2}\cdot\frac{c^2}{m^2 T^2}$$

$$\dots\dots(24)$$

By measuring the appropriate standard errors for the required variables the error in t can be calculated using the most suitable of the above equations (21), (22) and (24). Although the equations look rather complicated they are only a series of summations and hence are quite easy to use.

5.2 Monte Carlo Analysis

An alternative to the above method for estimating errors is to use a Monte Carlo analysis. Monte Carlo computer programs evaluate equations many times over with randomised input parameters chosen from a predetermined distribution. The technique can be used to solve problems of random walk, numerical evaluation of differential equations and convolution of distributions. The present problem falls into the last category.

The programs used in this type of analysis simulate the errors in an oxide dating calculation by means of repeated evaluation of equations (3) and (21). The population of results generated represents the error in the calculation if the population of the randomised values fed into the equations represents the errors in each of the input parameters x, m, G, K and T. The random number routine used generates a Gaussian distribution of values of given mean and standard deviation. Having determined the errors on the input parameters to the oxide dating equations, a series of random values of required mean and standard deviation are used to generate the required population of oxidation times.

Care must be taken in choosing the subroutine for random number generation. For example on an IBM370 computer random numbers can be selected using the internal GRAND subroutine or the IBM Scientific Subroutine Package GAUSS. The latter routine repeats at $2^{29}\pi$ with appropriate choice of seed and is more accurate than the former. The two routines differ by 0.37% in their calculation of mean t and 5.4% in σ_t after 1000 calculations for a particular data set where a statistical error of about 3% in σ_t would be expected from a sample of 1000. The Monte Carlo technique has sometimes been used on microcomputers but this has the problems of slow operation and unsuitable pseudo random number sequence generator programs. The subprograms supplied often give number sequences which inadequately satisfy the tests of randomness and repeat after quite short sequences. Careful investigation of sequences available should be made before carrying out Monte Carlo calculations on any computer.

5.3 **Examples of Analysis and Errors**

Let us now consider the application of oxide dating using a particular set of air oxidation data. For the alloy considered, 1%Cr 1%Mo 0.75%V, air oxidation data have been obtained by United Steel[10] and CERL[11]. For these examples we will use CASE 2 of Section 3.2 (i.e. single sets of data existing at different temperatures with extrapolation /interpolation required to a particular temperature of interest). Table 1 shows a set of oxidation thickness measurements taken from single specimens oxidised at temperatures between 500 and 600°C for times up to 10,000 h. For each data set at a particular temperature ln x versus ln t (thickness in μm and time in seconds) can be plotted (Figure 19) and the mean values of m and K are obtained along with their standard deviations. The values of ln K obtained for each data set are plotted against (1/T - 1/T) which gives values of K_1, σ_K, G and σ_G (Figure 20). This gives all the required terms for the evaluation of equation (22) and they are shown in Table 2. Two examples of their use will now be given.

Example 1

A component has been in service at a temperature of 520°C ± 4°C and during this time an oxide layer of thickness 134 ± 5 μm has formed. How long has the component been in service?

From equations (3) and (21) the total exposure time

163

is 49,145 h. The error on this value is given by substitution into equation (22);

$$\frac{\sigma_t^2}{t^2} = \underset{[\sigma_x]}{0.00449} + \underset{[\sigma_m]}{0.029} + \underset{[\sigma_{K_1}]}{0.0032} + \underset{[\sigma_T]}{0.0211} + \underset{[\sigma_G]}{0.00247} = 0.0603$$

Therefore $\dfrac{\sigma_t}{t} = 0.245$

Thus the total exposure time is 49,145 $\pm$ 12,040 h
    It can be seen that the largest contribution to the total error on t comes from the terms involving σ_m and σ_T. σ_m is dependent on the quality of laboratory data but σ_T depends upon knowledge of the exposure temperature.
    The same data have been analysed by the Monte Carlo technique described in Section 5.2. Figure 21 shows the distribution of calculated exposure time for errors in all the variables (Figure 21a) and also for errors in single variables (Figure 21b - f). The distribution is skew because of the exponential terms in the equation and a plot against ln(time) indicates a log-normal distribution (Figure 22). Both Figures 21 and 22 confirm that the greatest contribution to the error in the final result arises from errors in m and T. If the error on oxide thickness was doubled then the error on t rises from $\sigma_t/t = 0.245$ to $\sigma_t/t = 0.272$. However a doubling of the error on temperature causes the error on t to rise from $\sigma_t/t = 0.245$ to $\sigma_t/t = 0.352$ and a doubling of the error on m results in a value for σ_t/t of 0.384.

Example 2

    In the component described in Example 1 a wide crack is found and the oxide on the crack face is of uniform thickness of 50 $\pm$ 3 μm. Therefore the crack is assumed to be stationary.
    How long has the crack been present?

    From equations (3) and (21) the exposure time of the crack face is 8372 h. Determination of the error on this value is given by substitution with equation (22);

164

$$\frac{\sigma_t^2}{t^2} = \underset{[\sigma_x]}{0.0116} + \underset{[\sigma_m]}{0.0239} + \underset{[\sigma_{K_1}]}{0.0032} + \underset{[\sigma_T]}{0.0211} + \underset{[\sigma_G]}{0.00247} = 0.06227$$

Therefore $\dfrac{\sigma_t}{t} = 0.249$

Thus the total exposure time of the crack is 8372 h with a standard error of 2089 h.

Figure 23 shows the distribution of the calculated exposure time for the component and for the crack and it is evident that there is no overlap in the distributions.

It is of interest to calculate the crack age at which the oxide thickness distribution would become statistically indistinguishable from that calculated from the service life of the component, i.e. the crack formed before or immediately upon the component being put into high temperature service. For the data in this example a standard t-test shows that there is a 5% probability that the crack is as old as the component when the oxide on the crack face is 120 μm thick if it has a standard error of 5 μm (the exposure time of the crack is then calculated to be 40,312 h with a standard error of 9936 h).

It is clear from the forgoing analyses and examples that the oxide dating technique will provide useful results provided that the laboratory and plant data are collected properly. There are two aspects to the good application of the method. The first is good laboratory data collection and analysis, and the second is accurate characterisation of the service component under investigation. The variables which have the greatest influence on the accuracy of the result are m and T.

When laboratory data are being collected the temperature is usually fixed and well controlled (poor control on laboratory temperature will result in poor accuracy for m). It is therefore important to collect enough data for m to be determined as accurately as possible. It should be noted that the assumption that m = 0.5 in all cases can lead to significant errors. During the characterization of the service component it is often found that the exposure temperature is not known with any great accuracy, yet this is an important plant variable and has an important effect on the accuracy of the final result. One might even argue that more effort should be put into measuring the thermal history of the component than into measuring the oxide thickness accurately!

A further complication is that during the exposure life of the component it may have operated at more than one temperature. If this is the case then the data must be analysed using the method described in Section 3.1. Because of the possible complexity of the equations used in this method an analysis of the errors is more easily carried out using the Monte Carlo technique.

Two methods of analysing the oxidation data have been presented. Table 3 shows a comparison of the errors in the oxidation equation variables calculated by the Monte Carlo method applied to equations (3) and (21) and by the analytical method applied to equation (22). The results are in good agreement which indicates that ignoring any possible correlation terms in equation (18) has not led to errors in the analysis.

6. CONCLUDING REMARKS

A careful examination of the oxide covering the free and defect surfaces of high temperature components, coupled with a detailed knowledge of the relationship between oxide properties and service environment, may provide valuable information regarding the thermal history of a failed component. Oxide thickness measurements permit the estimation of exposure time or temperature, whilst oxide morphology and lattice parameter provide a means by which the exposure temperature may be estimated. The errors which can arise in the application of the oxide dating technique can be minimised by careful attention to the accuracy of the plant exposure temperatures and the exponent in the basic oxidation equation $x = Kt^m$.

7. ACKNOWLEDGEMENTS

This Chapter is published by permission of the Central Electricity Generating Board.

8. REFERENCES

1. L.W. PINDER: Corrosion Science, 1981, <u>21</u>, 749-763.
2. E. METCALFE, M.R. TAYLOR and J.P. BROOMFIELD: Corrosion Science, 1984, <u>24</u>, 871-884.
3. O. KUBASCHEWSKI and B.E. HOPKINS: 'Oxidation of Metals and Alloys' (Revised 2nd Edition), 1967, 108, Butterworths, London.
4. L.W. PINDER: 18th Corrosion Science Symposium, 1977, University of Manchester.
5. J.E. FORREST and P.S. BELL: 'Corrosion and Mechanical Stress at High temperatures' (eds V. Guttmann and M. Merz), 1981, 339-345, Applied Science Pub., London.
6. E.C. POTTER and G.M.W. MANN: 'First Int. Congress on Metallic Corrosion', 1961, 417, Butterworths, London.
7. M.I. MANNING and E. METCALFE:' Proc. BNES Conf. on Ferritic Steels for Fast Reactor Steam Generators' 1977, 378-381, London.
8. A. HOAKSEY, F. HICKS, P.C. ROWLANDS, and D.R. HOLMES: '6th European Congress on Metallic Corrosion' 1977, 37-45, Soc. Chem. Ind., London.
9. P.R. BEVINGTON: 'Data Reduction and Error Analysis for the Physical Sciences', 1969, McGraw Hill, New York.
10. United Steel Companies Ltd. Durehete 1055 Data Sheet Ref. No. USC 582.2.5.4.66, 1966.
11. E. METCALFE and M.R. TAYLOR: 1984, Unpublished Results.

TABLE 1. OXIDE THICKNESS MEASUREMENTS (μm) AFTER AIR
OXIDATION OF CrMoV STEEL

Temp (K)	Time (h)					
	1000	2000	4000	6000	8000	10,000
773	8.9	13.1	18.7	23.7	28.3	31.6
798	17.5	26.1	37.8	47.3	55.9	63.6
823	28.2	41.5	62.4	77.1	91.3	101.9
848	41.4	58.1	88.1	108.4	131.1	147.3
873	75.6	111.8	165.4	205.8	241.8	276.9

TABLE 2. MEAN VALUES AND STANDARD ERRORS OF THE TERMS
DERIVED FROM THE DATA IN TABLE 1

	Mean value	Standard error
m	0.557	0.005
K_1	0.00613	0.000194
G	12742	603.3
T	823	-

TABLE 3. COMPARISON OF THE ERRORS IN THE VARIABLES USED
IN THE OXIDE DATING CALCULATIONS BY THE
ANALYTICAL METHOD AND THE MONTE CARLO METHOD

Variables method	Analytical method	Monte Carlo method
	$(\sigma_t/t)^2$	$(\sigma_t/t)^2$
used in example 1		
All variables	0.0603	0.0600
x	0.0045	0.0044
m	0.0290	0.0279
K_1	0.0032	0.0035
G	0.0025	0.0024
T_1	0.0211	0.0219

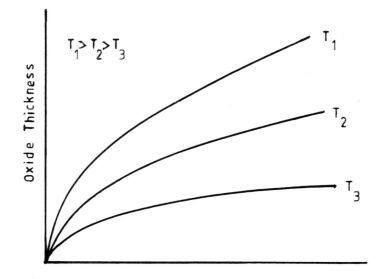

$T_1 > T_2 > T_3$

Oxide Thickness

T_1

T_2

T_3

Time

Fig. 1   IDEALISED PARABOLIC PLOT OF OXIDE THICKNESS AS A FUNCTION OF TIME

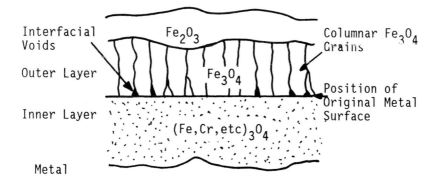

Interfacial Voids

Outer Layer

Inner Layer

Metal

Fe_2O_3

Fe_3O_4

$(Fe,Cr,etc)_3O_4$

Columnar Fe_3O_4 Grains

Position of Original Metal Surface

Fig. 2   SCHEMATIC DIAGRAM OF A DUPLEX OXIDE SCALE FORMED ON STEEL

Fig. 3   OPTICAL MICROGRAPH OF SCALE FORMED ON 9%Cr1%Mo
          STEEL IN STEAM

Fig. 4   SCANNING ELECTRON MICROGRAPH OF THE INNER
LAYER/OUTER LAYER INTERFACE OF THE SCALE
SHOWN IN Fig. 3

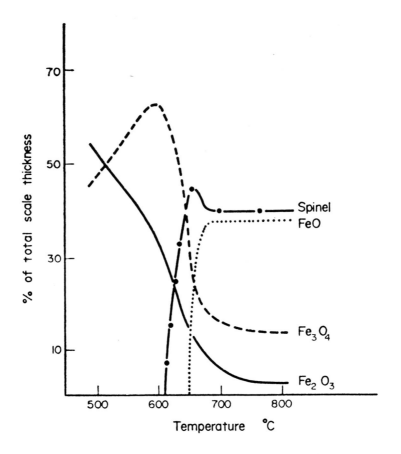

Fig. 5   VARIATION OF SCALE COMPOSITION WITH
TEMPERATURE FOR AIR OXIDATION OF 5%Cr1%Mo
STEEL[1]

172

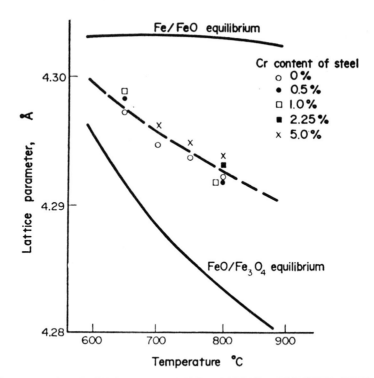

Fig. 6   VARIATION OF WUSTITE LATTICE PARAMETER WITH
TEMPERATURE[1]

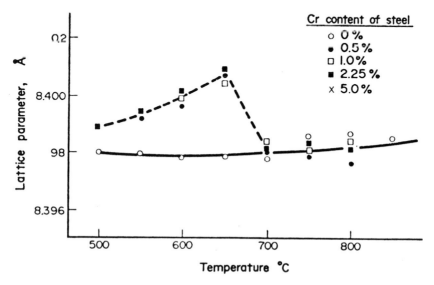

Fig. 7   LATTICE PARAMETER OF MAGNETITE AS A FUNCTION
OF FORMATION TEMPERATURE[1]

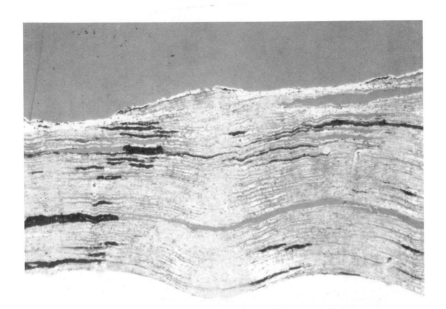

25 μm

Fig. 8   THE MORPHOLOGY OF LAMINATED SCALE FORMED ON
THE STEAMSIDE OF LOW ALLOY STEEL

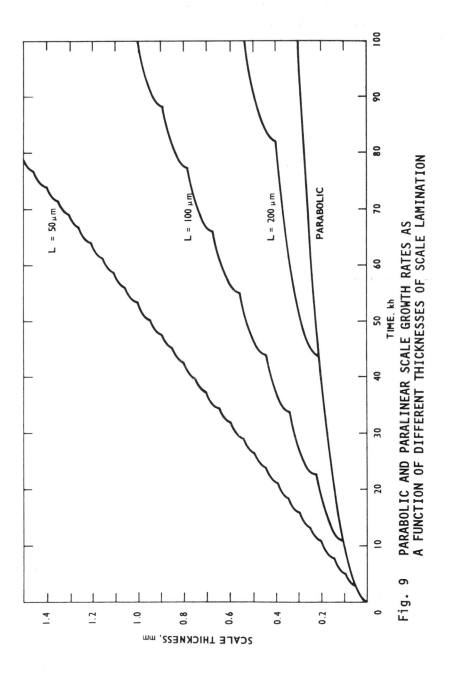

Fig. 9   PARABOLIC AND PARALINEAR SCALE GROWTH RATES AS
A FUNCTION OF DIFFERENT THICKNESSES OF SCALE LAMINATION

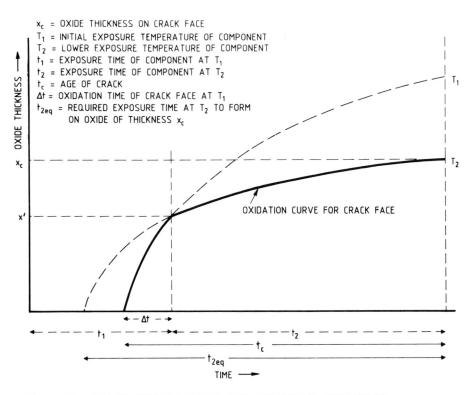

Fig. 10 CRACK OXIDE DATING FOR MULTIPLE OXIDATION TEMPERATURES

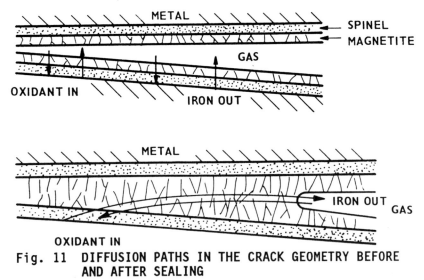

Fig. 11 DIFFUSION PATHS IN THE CRACK GEOMETRY BEFORE AND AFTER SEALING

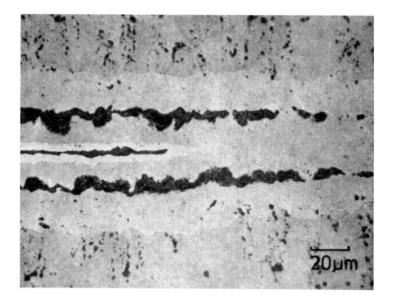

Fig. 12 HAEMATITE REDUCTION FOLLOWING BLOCKAGE OF CRACK IN MILD STEEL[1]

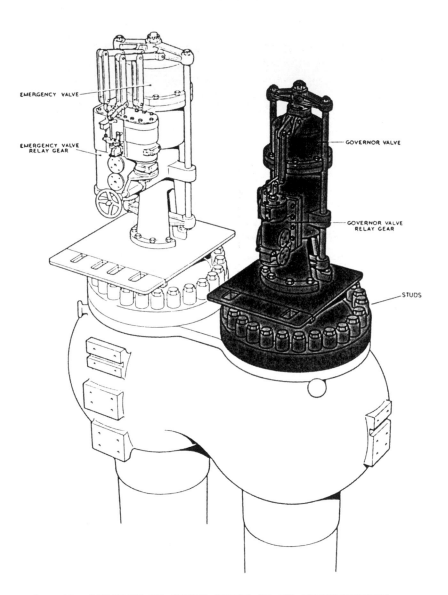

Fig. 13   DIAGRAM OF STEAM CHEST OF AN INTERMEDIATE
          PRESSURE TURBINE

Fig. 14   STEAM CHEST AFTER FAILURE OF GOVERNOR VALVE
RETAINING STUDS

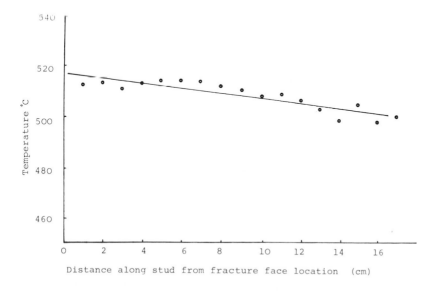

Fig. 15   TEMPERATURE PROFILE ALONG A STUD

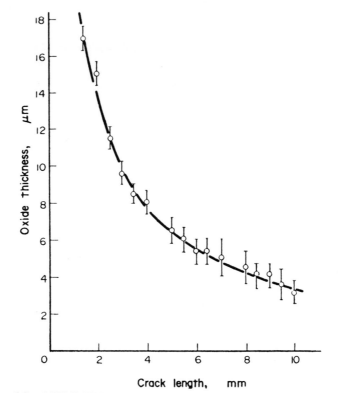

Fig. 16 OXIDE THICKNESS AS A FUNCTION OF CRACK LENGTH
FOR A DEFECT IN A C-Mn STEEL PIPE

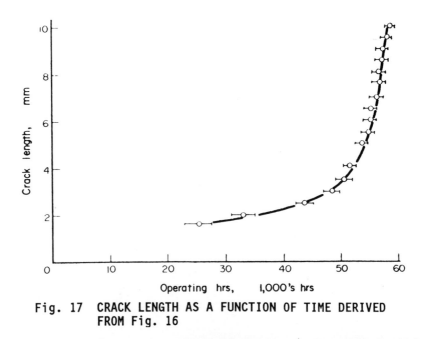

Fig. 17 CRACK LENGTH AS A FUNCTION OF TIME DERIVED
FROM Fig. 16

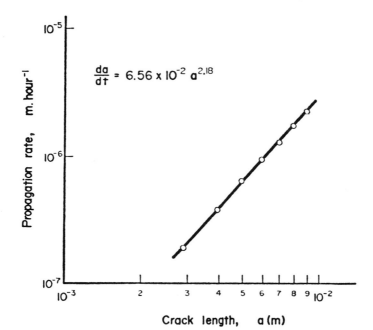

$$\frac{da}{dt} = 6.56 \times 10^{-2} \, a^{2.18}$$

Fig. 18   CRACK PROPAGATION RATE OF THE DEFECT DERIVED
FROM Fig. 17

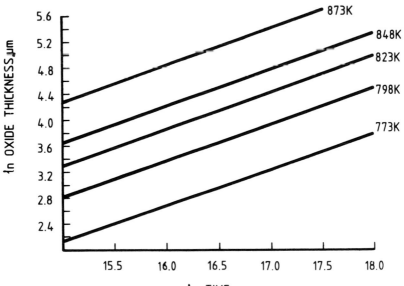

Fig. 19   LN(OXIDE THICKNESS) VERSUS LN(TIME) FOR CrMoV
STEEL AT VARIOUS TEMPERATURES IN AIR

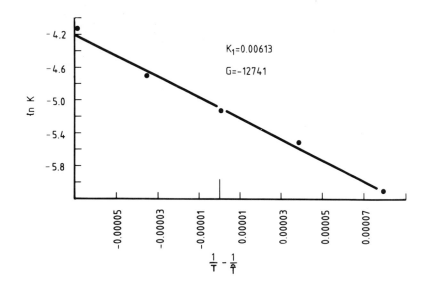

Fig. 20   PLOT OF K VERSUS (1/T - 1/T̄) FOR CrMoV STEEL
OXIDIZED IN AIR

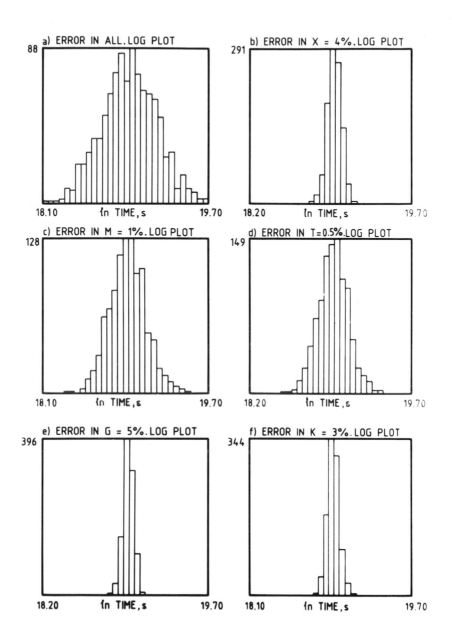

Fig. 21 DISTRIBUTION IN CALCULATED EXPOSURE TIME IN
OXIDE DATING CALCULATION REFLECTING THE
EFFECTS OF ERRORS IN a) ALL VARIABLES, b) x,
c) m, d) T, e) G, AND f) K

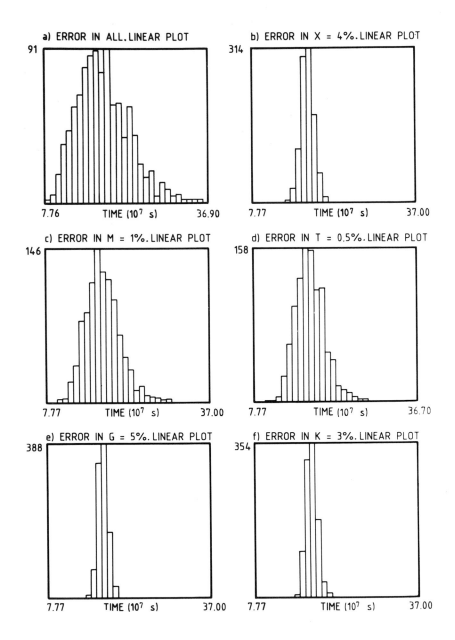

Fig. 22   DISTRIBUTION OF LN(CALCULATED EXPOSURE TIME)
IN OXIDE DATING CALCULATION REFLECTING THE
ERRORS IN a) ALL VARIABLES, b) x, c) m, d) T,
e) G, AND f) K

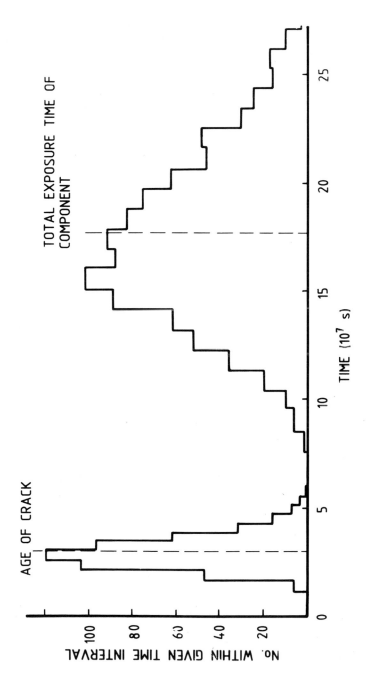

Fig. 23   DISTRIBUTION OF CALCULATED EXPOSURE TIME FOR
THE COMPONENT AND FOR THE CRACK IN THE
COMPONENT

5:The testing and evaluation of materials in tribology

T S EYRE AND F A DAVIS

There are a very large number of well documented test procedures for use in the determination of the tribological characteristics of engineering materials. Unfortunately there is little correlation between different tests or with service behaviour and the vast amount of information available is therefore difficult to use in engineering design. Friction and wear are not intrinsic properties of materials but are related to the 'engineering system in which they are used. In simulation tests it is necessary to establish the correlation between the laboratory test and service behaviour and this can be achieved by identifying the wear mechanism(s) operating.

To improve technology transfer and to compare new engineering wear resistant materials with more traditional materials it is necessary to have well established test procedures and data bases. A few tests, in particular the pin on disc are sufficiently well established and widely used to form a basis for standard tests and an established methodology. This can be achieved and will eventually lead to international standards, but only when the test(s) itself, the material(s) and the tribological environment are thoroughly understood.

1. INTRODUCTION

There are many well established methods for the determination of the physical, chemical and mechanical properties of materials and this makes it relatively easy to incorporate them into specifications. Specifications are used to establish quality assurance on the one hand and to characterise materials on the other, rarely do they attempt to correlate the material properties directly with service behaviour, although in a few cases design codes of practice are available for this purpose. Test procedures for hardness and other mechanical properties are fairly well established and engineers are familiar with their use and quite often material selection is based upon this information, combined wtih previous experience.

Unfortunately there is no direct correlation between the physical, mechanical and chemical properties of materials and their friction and wear behaviour although hardness is the most frequently used property to represent abrasion resistance and therefore forms a basis for materials selection. The hardness of the worn surface is a better measure of wear resistance than the unworn hardness, some alloys only develop their maximum wear resistance in service due to surface deformation or phase transformation and this only occurs if the surface stress is high enough.

The hardness of the wear resistant surface may vary significantly in three directions (1) and some surface coatings will vary through the thickness whilst others are relatively constant. A distinction must also be made between macro and micro hardness. As most materials including ceramics soften as the temperature rises it may be important to use the hardness generated at the operating temperature as a measure of wear resistance.

Any discussion of the hardness of the surface being worn cannot however be divorced from the hardness of the abrasive particles, Richardson (2) has shown that it is only when the ratio:-

$$\frac{\text{Hardness of the surface}}{\text{Hardness of the abrasive}} > 0.6$$

that significant improvements in wear resistance occur

(Fig. 1).   There is little to be gained by increasing this ratio much above 1.2 and it is questionable therefore what benefits occur with materials having a hardness greater than 2000 HV.   The hardness of many minerals and materials used for wear resistant purposes are contained in Table 1 and Fig. 2 and reference to these will enable the hardness ratio to be optimised.

This situation is even more complex however because there is little direct correlation between friction and wear as indicated by the values in Table 2, small variations often occur in friction for large variations in wear.   Generally speaking any increase in friction results in an increase in wear and the converse is also true, but the correlation is rather general, not quantitative and there are many anomalies.   Both lead and polytetraflorethylene are very good examples of materials with a very low coefficient of friction which also have a high wear rate and they cannot therefore be used in the bulk form.   They are however incorporated into a wide variety of polymers and composites with additional strength and increased wear resistance.

Friction arises through surface asperity interactions which involve mechanical, chemical and physical processes, but wear usually occurs as a result of fracture in the bulk of the asperity some distance away from the friction surface and it is not too surprising therefore that they are not directly related.

Friction and wear are not intrinsic properties of materials but are characteristics of the engineering system in which they are used (Fig. 3).   Reducing friction and increasing wear life may be brought about by a number of factors, some of which do not require a change in the counterface materials.   For example controlling the operating environment by improving filtration in an internal combustion engine may reduce wear by reducing the amount of abrasive particles in the recirculating oil.   Improving the surface geometry and/or surface finish may also bring about improvements which are a direct result of improving the supply of lubricant to the real contact areas and reducing the particle size of the wear debris produced during running in. (3)

Terminology in tribology still causes considerable confusion and this is best illustrated by the use of such terms as low stress and high stress abrasion, gouging and scratching, all of which are different forms of abrasion. The wider adoption of an OECD recommendation (4) has brought about some improvements, although greater efforts are still required to adopt standard terminology so that any dialogue between investigator and engineer can be carried out in a precise manner free from ambiguities. In this way technology transfer from research to industry is likely to be considerably improved and in this respect tribology is significantly behind other areas of engineering in supplying aids to designers. (5)

There is considerable current interest in tribology test procedures, particularly wear testing of materials, which has been brought about by an increased awareness of the economic and technological improvements in materials including composites, ceramics and surface coatings.

Industry has problems in identifying the optimum material to be used for engineering components, there are frequent day-to-day problems in repair and maintenance and there is much discussion of simulation and the relevance of wear data. In the development of new or improved materials it is necessary to collect relevant data and to compare and contrast them with more traditional materials and it is this aspect in particular which focusses attention on the methods available for wear testing. Improved testing techniques will assist tribologists and design engineers to collaborate through their common aim to select materials, optimise surface geometry and lubrication methods so that a given product life can be achieved.

Over the past twenty years since its introduction by the Jost Committee, (6) the word Tribology has become accepted and widely used throughout the world. There is now a move towards the use of "Surface Engineering" as an umbrella under which the production of surfaces, their characterisation and their behaviour in service can be drawn together through an interdisciplinary approach and to channel this in the direction of technology transfer.

2. FRICTION, WEAR AND LUBRICATION TESTING

Wear means different things to different people. On the one hand in abrasive processes it might be economically necessary to accept a 30% weight loss before the component is replaced, for example a digger tooth. On the other hand a piston ring for example may be inefficient with only 1 or 2 per cent of weight loss, where more importantly it is the change in surface topography which limits its life or efficiency. In the case of dies for polymer processing the die is worn out when approximately 50 um is worn off its working surface with the loss of fidelity of its textured surface. In the Railway industry uncontrolled friction often caused by leaves shed by trees in Autumn leads to wheel spin, with consequential damage both to wheel and the rail.

In many wear tests the frictional characteristics are completely ignored and the production of an acceptable low wear may be nullified by a high friction surface being produced which reduces the flow characteristics in bulk materials handling.

Wear, therefore may involve excessive weight or volume loss, an unacceptable change in surface topograhy or a change in shape. All of these factors must at least be considered before specific wear tests are discussed. There may be more than one wear mechanism encountered in plant and equipment. The synergism between wear and corrosion is now recognised as an important factor and greater corrosion resistance may be required in materials used for example in corrosive mining applications (7). During use the wear mechanisms may change because of the dynamic situation which exists, for example adhesive wear may change to abrasive wear because of the production of a hard wear debris. Cause and effect therefore must be separated and in the example just referred to the solution is not to make the materials 'harder', but to prevent adhesive wear initiating abrasion.

Friction, wear and lubrication mechanisms have been fully discussed elsewhere (8, 9, 10, 11) but it is appropriate to briefly discuss the main wear mechanisms that will be encountered in wear testing here at this point as a preliminary to wear testing.

2.1 Wear Mechanisms

Abrasion is caused by the penetration and passage of hard particles across the surface resisting wear, the result of which is removal or displacement of material by deformation, ploughing and cutting (Fig. 4a) with the production of spiral wear debris. Adhesion occurs when two surfaces rub together and asperities interact by adhesion and bonding to produce a strong weld. Continued motion produces deformation and fracture in the weaker of the two counterface materials resulting in transfer of the weaker to the stronger surface (Fig. 4b). Adhesive wear is often referred to as metallic wear due to its metallic appearance in contrast to oxidative wear (Fig. 4c) when the surface becomes discoloured by a thick oxide film. Transitions from metallic to oxidative wear are often encountered and these are also referred to as severe (metallic) and mild (oxidative) respectively. Another form of wear often encountered when abrasion and adhesion do not occur is delamination. This occurs by plastic deformation and fracture to produce a platelike debris (Fig 4d). Fretting occurs by an oscillating movement over a very small amplitude between surfaces in the presence of an oxiding atmosphere. Two significant features of fretting are adhesion of micro asperities in conjunction with oxidation to produce a small particulate oxide debris. Many wear situations consist of repeated stress cycles which cause a fatigue failure and the classic striations showing the stepwise nature of fracture is the main feature of the fracture surface (Fig. 4e).

There is a natural tendency to accelerate the wear test to accumulate data more rapidly. In increasing the load and speed in the test machine, care is required to ensure that the conditions produced are not so peculiar to the test itself that they are not related to service behaviour and therefore a correlation is not obtained. (12)

The Society of Tribology and Lubrication Engineers (STLE) documents some 300 different methods of friction, wear and lubrication testing, and these are broken down into about twelve different categories (13) (Fig 5). The most frequently used of these devices are shown in (Fig. 6) and of these the pin on disc is most frequently referred to in current tribology literature. Two other very well known tests have been standardised through the American Society for Testing of Metals (ASTM).

Four-ball test   ASTM D2783-71, for lubricant testing

Rubber Wheel test ASTM G65-81-A for abrasion testing

For these tests there are also standard reference materials, namely A.1.S.1.55100 steel balls for the former and a National Bureau of Standards steel designated "1957" for the latter. The message to be learned through the efforts of ASTM is that all of the test parameters must be systematically investigated to enable a reproducible test to be carried out and this has been achieved with the Rubber Wheel Test, which has now been widely adopted by industry in the USA and elsewhere for quality assurance in the supply of abrasion resistance metals and alloys. One of the most critical factors in this test however is the quality of the rubber wheel and in the USA it is said that only one manufacturer produces an acceptable product. The other problem with this test is that a large quantity of standard abrasive particles are required for each test and the test itself is not very suitable for development of materials for applications which do not fall into the category of low stress abrasion. The problem of friction measurement must also be addressed because it affects the flowability of the material processed and in this respect some commercial machines including the rubber wheel test do not measure friction.

Although there are hundreds of different test methods and these are still increasing in number, the reasons for testing are relatively few and may be summarised as:-

1. To obtain fundamental information on mechanisms. It is usually necessary to simplify the test procedure, by running at low loads and speeds and in a controlled atmosphere and utilising pure materials.

2. To characterise materials and to determine the effect of variations in composition, micro-structure and mechanical properties on

192

tribological behaviour.

3. To simulate service behaviour where it is generally felt necessary to reproduce all of the variables experienced in service in the simulative test otherwise misleading information is obtained and incorrect decisions may be made.

Of these, type 3 is the most difficult because of the complexity of simulation, the tendency being to either miss out or not be aware of some important variables or if all factors are taken into account the simulation machine ends up as a copy of the service system, i.e. an engine to simulate an engine or a jaw crusher to simulate a jaw crusher. Of the hundreds of tests available most fall into category three because of the need of industry to simulate field conditions and to improve their technology. Great care is required to obtain a satisfactory correlation between the laboratory test and field behaviour. All too often a poor correlation is achieved and confidence is lost in that particular test method. It is absolutely necessary to reproduce the wear mechanisms observed in the service environment in the laboratory test, otherwise the results may be misleading. A wide rage of diagnostic techniques therefore are used to examine the surface topography, its structure and composition and to identify the morphology of the debris and to relate it to the wear surface so that wear mechanisms can be identified. Some of the more widely used important techniques are shown in Table 3.

3. GENERAL CONSIDERATIONS IN WEAR TESTING

The first and perhaps most obvious difficulty in practical wear testing is the range of wear rates which may be encountered. Mild steel rubbing on itself wears at 10 times the rate of a tungsten carbide-cobalt alloy rubbing on itself under similar conditions (14). In the case of steel, wear rates beteen 10-11 to 10-5 cm^3 are encountered for loads in the range 0.20 N to 200N for speeds of 50-200 cm/s (15) (Table 4). Thus there are dificulties in designing a machine of adequate versatility, either for testing a variety of materials or even just covering the complete range for one material. Sufficient sensitivity must be incorporated to measure wear rates at low loads while remaining robust enough to support loads at the other end of the scale.

The vibration developed in rubbing, particularly at high loads and speeds, is a particularly severe limitation in the design of a wear test machine. Sensitive measuring equipment may be unable to function under severe vibration, and thus may only be used at low speeds and loads, employing more robust forms of measuring equipment under more arduous conditions. (16)

The heat generated in rubbing is another important consideration. The temperature may rise beyond tolerable limits for measuring equipment, strain gauges and linear displacement transducers being particularly vulnerable in this respect. There is, however, some scope for control here in the use of insulating materials, and in some cases it might even be advantageous to employ some means of refrigeration.

On these three considerations alone, it does not seem possible to successfully design a wear test machine which will successfully deal with the whole range of engineering materials and conditions met with in service. These objectives may only be achieved by the use of two or more machines, resembling each other as closely as possible (e.g., use of similar specimens, method of load application) and differing only in their robustness of construction and the associated instrumentation.

3.1 Classification of Problems Encountered

It is possible to classify the difficulties encountered in designing a wear testing system in the following way:

(a) Problems associated with the application of load and relative motion.

(b) Problems associated with specimen size and shape.

(c) Problems associated with specimen condition.

(d) Problems associated with environmental
    conditions.

3.2 Load and Speed

Problems associated with the application of load and
sliding movement to the test specimens usually revolve
around relative costs and convenience in operation. The
load may be applied hydraulically, pneumatically, by
weights, springs, or even by electro-mechanical devices,
as required. The only really important requirement is
the accuracy of measurement of the load and that this
load remains at a constant value throughout the test.
Virtually any prime mover may be used to impart relative
motion provided it gives enough torque, and the sliding
speed remains constant. It is also necessary to be able
to vary the sliding speed and to be able to measure the
speed accurately.

3.3 Specimen Size and Shape

The size of specimens must be such that:-

1.   There is sufficient material available to allow the
     test to be completed;

2.   The specimens do not distort under the influence of
     the applied load and the sliding motion.

     It is desirable, if dimensional changes are used as
a basis for wear measurement, that the specimens should
be shaped so that the removal of a small amount of
material causes a relatively large dimensional change to
at least one of them (e.g. 'pointing' one specimen which
is rubbed against the flat surface of another).
However, if high wear rates are encountered, such
shaping may cause the material to be worn away too
rapidly, and in view of the range of wear rates
previously referred to, some compromise must be arrived
at.

     It might be thought suitable to use specimens of
different size and geometry for low and high wear rate
tests. However, a change in specimen size and geometry

is likely to cause a change in the temperature generated at the sliding interface, with the possibility of a change in the nature of the wear mechanism involved. Two factors influence the interface temperature (17), the heat produced by friction and the cooling characteristics of the specimens (the cooling rate of the specimens, at least in dry wear studies, is not likely to change significantly unless the environment is changed).

The temperature rise at the sliding interface (usually referred to as the 'flash temperature' rise) above the bulk temperature of specimens, due to asperity contacts, is independent of specimen size and can be estimated fairly accurately (18). The bulk temperature of the specimens will depend on the rate of heat loss by the specimens to the surroundings, and this in turn depends on the specimen size and geometry. The bulk temperature rise of the specimens is not easily calculated, and usually if this is to be found, measurements are made.

A further point can be made concerning specimen size and shape. In certain test arrangements, e.g. when one specimen in the form of a pin is rubbed against the flat surface of another in the form of a disc spinning at constant speed, the finite area of the wear surface of the pin means that the sliding speed will vary across the surface in a direction perpendicular to the sliding motion. The dimension of the specimen in this direction should be as small as possible. Furthermore, if the specimen shape is such that this dimension changes during the progress of wear, the speed variation across the surface becomes more marked with the progress of time. Whenever possible, this design feature should be avoided.

3.4 Condition of Specimens

The factors to be considered include surface roughness, surface chemical composition, the microstructure and the mechanical properties of the material which can be related to composition. Wear which occurs during testing can, and usually does, alter the state of the surface and subsurface layers. Surface roughening, change of microstructure and hardness and generation of surface films are examples of changes which may arise as a result of wear. These changes are time-dependent, and impose some limitations on test machine design. Firstly, provisions must be made for the control of environmental factors, e.g. humidity, atmosphere etc., as these factors frequently affect the course of these changes. This will be considered in more detail later. Secondly, the progress of these changes, and indeed whether they occur or not, will depend on the design of machine. We can differentiate two broad categories based on whether or not equilibrium wear conditions can be attained. 'Equilibrium wear' means that all major changes which can occur in surface roughness, hardness, surface films etc., have been undergone and the wear rate has become constant. The categories are:-

1. Abrasion occurs in such a way that fresh, unworn material is continuously exposed to rubbing, as a result, running-in is not attained. Examples of this are found in Archard's crossed-cylinder machine (19) and other variations, or the arrangement of Sata (20) where a lathe tool placed ahead of the wear specimen rubbing on a flat surface continuously removes worn material from the flat surface.

2. Sliding occurs over a fixed distance and is then repeated over this distance, i.e. sliding continues on the same wear track. As a result equilibrium conditions are attained provided the initial high-rate 'running-in' stage of wear is passed. The majority of wear test machines fall into this category, using either a reciprocating action or rotation of one or both specimens.

For a given material, tests conducted in the manner of the second category show that in a fairly short time, the surface roughness of the wearing surfaces reaches a constant value, this value again being dependent on the conditions of testing, which is independent of the initial surface roughness of the specimens (21). However, many investigators using this type of test arrangement have taken considerable trouble to prepare specimens with a standard, highly polished, initial surface; a procedure which would appear to be unnecessary. Only when changes in surface roughness during the progress of wear are being followed is it necessary to produce a highly polished initial finish, and in most other cases an as-ground finish should be sufficient, as the surface roughness generated will be considerably rougher than the ground surface.

For tests conducted in the manner of the first category this is not the case. Here, equilibrium wear conditions are not attained and the degree of asperity welding will depend, for a given set of test conditions and environmental factors, on the size and distribution of asperities, i.e. the surface finish. It is imperative that the initial surface roughness is known and as the degree of roughening is likely to be small, a really smooth surface is used.

Surface contamination of the specimens by grease or oil must be avoided. Even small amounts may cause a marked reduction of friction and wear. For a similar reason, metallic surfaces should be free of oxide. It is, of course, impossible to remove the last traces of oxide and adsorbed films without special treatment, but for wear testing in air at normal atmospheric pressures specimens should be given a degreasing treatment prior to testing.

There is much evidence to show that the micro-structural characteristics and hardness of the specimens markedly influence wear characteristics. The size and distribution of graphite flakes in grey cast iron is known to exert considerable influence on the wear rate and the load at which a transition from mild to severe wear takes place (22).

In steel and cast iron the presence of free ferrite is detrimental. The microstructure will have a considerable influence on the hardness, which may be critical in determining what mode of wear operates. It is imperative therefore that a constant and consistent source of material be available for test.

Changes in wear rate may arise as a result of the formation of hard surface layers during sliding, the hard layers being formed by phase-changes at locally-heated 'hot spots' (23). Variations in the microstructure can affect the kinetics of phase-changes, so it is conceivable that the microstructure could influence the extent of these changes, perhaps even to the point of complete suppression of certain types of wear.

3.5 Environmental conditions

In the design of a wear testing system, provision must be made for the control of the environmental conditions. For dry wear testing in air, by far the most important environmental variable is that of relative humidity. Not enough work has been done to make any generalisation on the effects of relative humidity on wear, but it has been shown that the abrasive wear of cast iron is at a minimum at 30% relative humidity (24) and that the type of oxides formed during the wear of wrought iron and mild steel is affected by relative humidity (25).

In many investigations it is helpful to consider the effects of atmospheres other than air, and possibly the effects of a vacuum environment as well. Variation of the temperature of the surroundings can also affect the formation of surface films. In a versatile wear testing system, some control over these variables is needed.

4. WEAR TEST MACHINE DESIGN

Examination of the literature on friction and wear shows that the test machines used differ considerably in detail, but the great majority possess a number of common features, although any one machine may not incorporate all of these. These features are:-

1. A prime mover, with or without means of varying the speed of sliding.

2. Load application device(s).

3. Specimens and their holders.

4. Environmental control.

5. Wear (and sometimes friction) measuring devices.

4.1 Prime Mover and Sliding Speed Control

In virtually all investigations some form of electric motor has been used to provide the relative motion between the specimens. This is usually the most convenient means of doing so. The only exception to this is when it is desired to use a very high sliding speed. It is difficult to set a practical limit to the sliding speed obtained from an electric motor, because the speed also depends upon specimen configuration, but a reasonably representative figure is about 3000 cm/sec which Johnson et al (26) achieved. Other methods for obtaining high sliding speeds (mainly in friction experiments) have involved the use of air turbines (27), a spinning ball in a vacuum suspended by a magnetic field (28), and a rocket-propelled sledge (29). For most purposes, these methods are at present of either academic or a highly specialist interest only.

It is desirable that the power transmission to the specimens should be capable of allowing controlled variation of sliding speeds. This may be achieved by a pulley and belt system using different sized pulleys, or a variable speed gear box, but the most satisfactory method is to use an infinitely variable speed drive. It is possible to obtain commercially a motor and variable speed drive as one integral unit; power take-off axis can be horizontal or vertical, and further reduction gearing can also be built into the unit if desired.

4.2 Application of load

The factors governing the configuration of the load application device include the load range required to be covered, the size and geometry of specimens, and the relative ease with which wear debris can be collected.

We may distinguish at least four configurations of load application device:

1. Load applied perpendicular to the axis of rotation, with a loading axis horizontal.

2. Load applied perpendicular to the axis rotation, with the loading axis vertical.

3.   Load applied parallel to the axis of rotation, with
      the loading axis horizontal.

4.   Load applied parallel to the axis of rotation, with
      the loading axis vertical.

These four configurations are illustrated schematically in (Fig. 7). In cases (1) and (2) a curved
wear surface is obtained, and we shall call these
arrangements pin/ring configurations. In cases (3) and
(4) a substantially-flat wear surface is produced, and
these will be called pin/disc configurations. It might
seem that some machines cannot be accommodated under
this classification; for instance, the various crossed-
cylinder arrangements such as those of Archard (15)
where two cylinders rotate while loaded against each
other, combined with axial movement so that helical wear
tracks are produced (Fig. 8). Here, a load (W) acts
vertically in a direction perpendicular to the axis of
rotation, so this is really a variation on configuration
(2). The axial movement is only to obtain wear
continuously over previously undisturbed material, and
this can be accommodated for, in all the configurations
shown in (Fig. 7) by providing for movement of the
loading axis such that either helical, cases (1) and (2)
or spiral, cases (3) and (4) wear tracks are produced.

The factors governing the choice of these
configurations involve several compromises between
conflicting requirements. Firstly, we need to know
whether flat wear surfaces are required, as given by
pin/disc configurations, or whether the curved wear
surfaces of the pin/ring systems are satisfactory.
Where microscopy of the wear surfaces is contemplated,
it is better to have flat surfaces, as curved surfaces
may give rise to depth-of-field problems, and taking of
replicas for electron microscopy becomes complicated.
Hardness measurements are not possible on a curved
surface unless the radius of curvature is very great
compared with the diameter of the impression.

Purely from the ease with which a known load may be
applied to the specimens, configurations (2) and (4),
where the load acts vertically downwards are much
superior to the others. In these cases the direct
application of a weight, perhaps through a simple lever

arrangement, is possible. However, while the removal of loose wear debris from the specimens in configuration (2) presents no particular problems (the debris merely falls away and can be collected) in configuration (4) some arrangements must be made for debris removal, e.g., by the use of scraper arms on brushes across the disc or the flat, (30) or the use of air jets playing on the wear surfaces of the disc, blowing loose debris away.

Debris removal in configurations (1) and (3) where the loading axis is horizontal, is no problem because the debris falls away. In these cases the orientation of the loading axis presents problems. It is still possible to use weights for loading the specimen if a rather complex system of levers and pulleys is used, but in such a system the frictional resistance is much more difficult to estimate and leads to some uncertainties as to the actual load on the specimen. Weight loading in these cases is also inelegant mechanically, and as an alternative hydraulic, pneumatic or spring loading can be resorted to (these methods can, of course, be used in the other configurations but for reasons to be discussed are considered inferior in these cases). Hydraulic or pneumatic systems appear at first sight to be the logical choice when the loading axis is horizontal, but again uncertainties in the frictional resistance of plungers, which may be variable, necessitate careful calibration with frequent checking. Furthermore, the frictional resistance in such devices can be so high as to rule out tests at low loads, unless another loading system to cover this range is resorted to. These objections also apply to some extent to springs and, in addition, if the pin wears away to a sufficient extent there will be an appreciable drop in load due to loss of spring compression.

Because of these conflicting requirements it is almost impossible to make any firm recommendations as to which configuration is the one to adopt for most applications. Individual requirements must decide this, but perhaps some generalisation can be tentatively made. Firstly, where a relatively simple machine is to be designed, capable of performing simple wear tests with the minimum of setting-up difficulties, but where only the wear rate is to be determined with a minimum of supporting observations, then pin/ring configuration (2) appears to be the best, using weights for specimen loading.

Secondly, where it is desired to have a test machine which will allow determination of the wear rate and supporting observations on specimens and the debris produced, either of the pin/disc configurations should prove satisfactory. Configuration (4) is inherently better than (3) for covering a wide load range (because weights are easily used) but is inferior from the point of view of debris removal. There is little reason in any instance for favouring configuration (1) but this has been included for completeness.

Finally it should be mentioned that a system which allows the load on the specimen to be indicated or recorded electronically, e.g. strain gauges and potentiometric recorder, is very desirable and goes much of the way to eliminating uncertainties in load on the specimens, particularly when hydraulic or pneumatic systems are used, as the load can usually be continuously monitored throughout the test.

4.3 Specimens and Specimen Holders

The principal factors which will affect the selection of specimen size and shape are:-

1.   The wear rate expected of the material under the whole range of loads and speeds to be studied;

2.   The methods by which wear is measured, i.e. whether by weight loss or dimensional changes.

Some estimates of the wear rate to be expected can usually be determined by examination of the relevant literature. If no data for a given material is available, then the only way out of the difficulty is to build a simple test machine to obtain such data. Luckily, for all common engineering materials sufficient data is available to enable at least rough estimates to be made, and for purposes of specimen design only the orders of magnitude of the range of wear rates is needed.

If wear is to be measured by following the loss of weight of specimens by removal of worn material, this method can be considered to be satisfactory only when 'steady-state' wear conditions are operative, as the specimen or specimens have to be removed for weighing. If changes in the wear rate are occurring, their progress may be modified by stopping the machine and the resultant cooling of specimens.

In other cases, and this is more usual, the method of measurement of wear is by following dimensional changes in the specimens occurring as a result of wear. While it becomes possible to measure wear while the specimens are running (and ideally this should be done) often the machine is stopped for the purpose of taking readings, usually because of vibration problems. If readings can be taken with the specimens in situ, or at least if they can be rapidly removed and replaced with the minimum of disturbance, then equilibrium conditions can be restored after running again in a reasonably short time.

In the case where dimensional changes are used as a basis for measurement, we can distinguish two classes:-

1. The apparent area of the wearing surfaces (as distinct from the real area of contact) remains constant with time;

2. The apparent area of the wearing surfaces increases with time.

Examples of wear pins, illustrating these two classes, for either pin/ring or pin/disc arrangements are shown in (Fig. 9). The dimension measured for evaluating the wear rate is l, and in (Fig. 9a), for steady-state wear conditions varies linearly with time as the apparent area of the wear surface is constant. In (Fig. 9b), for steady-state conditions varies with time in a manner dependent on specimen geometry.

These differences are important when wear is being recorded automatically. It becomes easier to follow the progress of wear when dimension varies linearly with time for any given steady-state. It is, of course, possible to obtain recording equipment which is able to make the necessary corrections even when it is not a linear function of time, provided specimen geometry is not too complex, but this adds to capital expense. Finally in this connection it should be mentioned that where dimensional changes are used for measuring wear, wear rates are usually expressed in terms of volume removed per unit sliding distance, so in any case there is still some conversion necessary in final presentation.

The main advantage in using pin geometries as exemplified by (Fig. 9b), is that the removal of a small volume of material causes a comparatively large change in dimension, effectively increasing the sensitivity of measurement. Obviously, if this is carried too far in some cases material may be removed at so high a rate that meaningful results are not obtained. For instance, mild wear is preceded by a period of high-rate severe wear, and it could turn out that the pin wears away completely before mild wear conditions have been established.

Even if this does not occur such a situation may still develop when a large load range needs to be studied, setting an upper limit to the load to be covered, as the differences in wear rates in mild and severe stages at high loads are great. The design of test piece geometry is thus a matter for compromise after careful consideration of material, range of loads and speeds to be studied, and other relevant factors, such as environment.

Only very general conclusions can be made concerning pin geometry, since so much depends on the actual material to be studied and the test conditions to be covered. For high loads, where high wear rates are likely to be encountered, pins such as those of (Fig. 9a) should be used. This will minimise rapid wearing away. For lower loads, pins of the type shown in (Fig. 9b) should be used, where greater sensitivity is required. Obviously, the increase in sensitivity depends on the apex angle at the front of the specimen. It should also be noted that the conically-shaped pin is, of the four, most unsuited to pin/disc configurations, because the radius increases so that the width of the wear track increases, and this may be too great to be tolerated. Finally, while it is possible to use one pin geometry for high loads and another for low loads so as to obtain the advantages of both, it must be emphasised that this can lead to possibly large changes in interface temperature, as explained earlier. Pin geometries should be as simple as possible so that realistic estimates of their heatflow characteristics may be made.

There is no special problem as far as discs or rings are concerned. The main consideration is that where possible the diameter of the disc (or the width of a ring) is made sufficiently large so as to allow more than one test to be carried out on its surface; in the case of a disc, both sides may be used also.

The design of specimen holders will affect the bulk temperature rise of the specimens; being in contact with the specimens the holder will affect their cooling characteristics. Perhaps more important is the fact that excessive heat may be transmitted from the specimens via the holders to other parts of the machine, and this will affect an enclosed environment, or prevent satisfactory operation of measuring devices. The use of insulating materials strategically placed will usually cure such problems.

4.4 Environmental Control

As mentioned previously, the environmental variables which should be controlled are relative humidity, nature of the surrounding atmosphere, and atmospheric temperature. The most important of these is relative humidity, and in most locations the day-to-day variation in humidity is significant enough to require the design of some sort of enclosure in which the specimens can rub under controlled humidity. In addition, such an enclosure can be designed so as to allow wear testing to be carried out in a variety of atmospheres and possibly in a vacuum. If only humidity control is sought, it is possible to use an enclosure that is not completely airtight; limited access of outside air will not significantly affect the humidity equilibrium attained in the enclosure, particularly when the enclosure is large. This reduces sealing problems, which must of course be considered in more detail when other atmospheres or high vacuum are contemplated.

For control of humidity, all that may be necessary is to place a saturated solution of a salt which contains excess solid phase inside the enclosure, and at any given temperature a definite relative humidity will be attained in the enclosure. Suitable salts can be chosen from the standard reference literature, and for convenience a few are listed in (Table 5) (31). One point to note in this connection, is that some salts give a wide variation in humidity for only a small temperature change; these should be avoided if possible.

In large enclosures the time required for equilibrium humidity to be attained after closure, may be excessive. One method of overcoming this is to blend dry and saturated air and pass the mixture through the enclosures. The time to attain equilibrium is then not only dependent on the size of the enclosure but also on the flow rate. By suitable adjustment of the flow rates of dry and wet air any desired relative humidity between 0% and 100% can be obtained (32). If the flow rates are fast, equilibrium is more quickly attained in the enclosure. The use of a recording hygrometer simplifies monitoring of the humidity level.

Some control of environmental temperature may be necessary, even when testing is carried out at what are equivalent to room temperatures. In small enclosures the frictional heating of the specimens may cause an appreciable temperature rise in the enclosure, and this may in turn lead to changes in humidity. If enlarging the enclosure does not minimise the problem sufficiently, then refrigeration will be necessary.

Heating of the environment to high temperatures is usually accomplished by surrounding the specimens with an electric furnace winding, or by the use of radiant heating bars. Occasionally specimens themselves can be heated by conduction through specimen holders. Thermocouples are used to monitor the temperature.

When tests are to be conducted in atmospheres other than air, the problem of sealing is foremost. Full use should be made of such devices as wilson seals, gimbals,

O-ring seals, bellows etc. A reciprocatng action rather than a rotary motion is sometimes favoured when conducting tests in controlled atmospheres. This can ease sealing problems as the reciprocating shaft connected to one of the specimens can be rigidly sealed and the reciprocation allowed for by bellows. Good examples of apparatus which successfully employs these principles are shown in the papers by Mitchell and Crawford (33) and Cornelius and Roberts (34). Occasionally in small machines using small loads it is possible to impart motion to the specicmens by magnetic coupling, eliminating the need for a shaft seal.

In recent years there has been an increasing interest in high temperature testing due largely to the possibility of using ceramics in advanced engines. Duffrane (35) has used a flat on flat reciprocating geometry to simulate the piston ring on cylinder liner configuration in the internal combustion engine with tests carried out up to 650 C. Matharu (36) has used a new high temperature device utilising a disc sliding against a disc in a atmosphere contolled chamber capable of temperatures up to 1500 C (Fig. 10). Tests carried out up to 900 C show that the coefficient of friction of RBSN* for example is much lower in air (0.20) than in vacuum (0.50) at temperatures below 700 C (Fig. 11). Wear data for the more widely discussed engineering ceramics show the superiority of silicon carbide over the whole temperature range with negligible wear even at 900 C (Fig. 12). This work shows the protective nature of oxides formed on the ceramic surfaces and the poor wear behaviour of the yttria partially stabilised zirconia.

4.5 Measurement of Friction and Wear

In many testing systems frictional force between the specimens is measured in addition to the wear of the specimens, as changes in wear mechanisms are usually reflected by changes in the coefficient of friction.

There are basically three ways in which specimen wear can be measured:-

---

*Reaction bonded silicon nitride

1.   By weight loss of the specimens,

2.   By weighing the wear debris produced,

3.   By following dimensional changes occurring as the
     result of wear.

Nothing more sophisticated than a chemical balance
is needed to carry out wear measurement by the first two
methods. The second has the advantage that wear can be
measured while the test continues. The chief
disadvantage of these methods is that they can only be
used when a loose debris is produced, and large errors
may be caused by wear debris adhering to the test piece.

If some of the wear debris remains attached to the
specimens, dimensional changes are more satisfactory.
For instance, if the rate of advance of a pin as it
wears against a disc specimen is measured, most of the
adherent debris is burred over at the edges of the wear
surface of the pin and does not significantly interfere
with the measurement. There are at least three ways in
which the rate of advance of a pin into a disc or ring
specimen can be followed, without the need for stopping
the machine (except where very accurate readings, free
of vibration, are required):-

1.   Mechanical methods, e.g. using a micrometer dial
     gauge;

2.   Optical methods, e.g. observing the movement of the
     pin by means of a travelling microscope or
     cathetometer;

3.   Electronic methods, using a linear displacement
     transducer coupled to the pin, and a potentiometric
     recorder to measure the change in resistance.

Electronic methods such as the example given are
probably the most accurate and convenient, but vibration
and heat development during running are serious
obstacles to be overcome. If the vibration caused by
rubbing is severe, then mechanical methods are likely to
be the most satisfactory.

It is desirable to have means of determining the coefficient of friction between the rubbing specimens. This can usually be accomplished in one of the followng ways:-

1. Fixing the stationary specimen to a rigid member such that the frictional force developed during sliding is exerted on this member. The strain arising in the stressed member can be measured using strain gauges. A variation of this is where the specimen holding assembly is freely suspended but rests on a load cell. The frictional force developed in sliding causes a downward force to be exerted on the load cell.

2. The stationary specimen holding assembly is freely suspended and deflects to an amount proportional to the frictional force. By previous calibration with weights, it is possible to estimate the frictional force developed in sliding.

3. By measurement of the torque loss in the motor which results from the friction between the two sliding surfaces.

In view of the problems to be overcome in the design of a practical wear testing system, and the need for compromise in achieving this, it is doubtful that any test rig can be designed to be of universal applicability. However, there are some areas in which it is possible to achieve greater standardisaton. For instance, the loading configuration described could be reduced. Thorough investigations involve the use of a variety of techniques for specimen examination, e.g. electron and optical metallography, electron probe microanalysis, x-ray and electron diffraction, microhardness measurements, surface roughness measurements, etc. As pin/ring configurations are not suitable in this respect, pin/disc arrangements are to be favoured as a basis for standardisation of load application. The development of hydraulic or pneumatic loading systems of sufficient accuracy would favour further standardisation to one configuration only., that of (Fig. 9b) where the pin is loaded horizontally on to the disc.

For similar reasons, it is not possible to design

standard test specimens for all materials under all conditions. Much depends on the material, and the test conditions. However. it is felt the design could be limited to two types of pin, run against discs. The pin of circular cross-section in (Fig. 9a), and the square-section pin of (Fig. 9b) could be the basis for such standardisation of pin size and geometry. Similar standardisation is possible for discs, a 10 cm diameter disc, 0.50 cm thickness, is convenient to handle and can allow several tests to be carried out with 0.50 cm diameter (or width) pin.

In many investigations, not enough attention has been paid to environmental conditions, particularly with respect to relative humidity. Some progress towards more standardisation in this connection is at present being made. A test machine should incorporate provisions for the control of relative humidity and surrounding temperature, by the use of a suitable enclosure for the specimens. Provision for testing in atmospheres other than air can be conveniently made in such an enclosure.

As experience in these techniques is gained it is becoming clear that it is advantageous to monitor friction, wear rate of the pin and temperature rise continuously and that intermittent recorders however fast their response time, may well miss important changes.

Microprocessors with a suitable software package are now available which can be coupled to the tribo test device.

There are many references in the literature where mild or severe forms of wear have been recorded, but from the information supplied about the sliding distance it would appear that insufficient time has been allowed for the material to completely run-in to a steady state condition. Even when high wear rates occur during the early stages of rubbing the investigator must resist the temptation to discontinue the test and record wear as severe. As the steady-state condition varies for different material combinations and, for a standard material varies also with both the load and the speed at which the test is performed, it becomes clear that no preconceived ideas about the sliding distance should be held by the investigator. Wilson (37) has discussed this aspect with particular respect to the wear of cast iron and it is not necessary to discuss this matter in further detail here.

4.6   Time or Sliding Distance of the Test

If the test is carried out for too short a period of time it will not be clear if all transitory effects have been examined. This is particularly so if the test data has not been plotted continuously. In metal to metal testing a high rate of wear initially reducing to a lower wear rate is quite common (Fig. 13). However in the case of an abrasion test it is likely that a test having a wear rate curve of the type shown in (Fig. 14) will give a very misleading result. Examination of the abrasion media, particularly if it is an abrasive disc will show that either clogging of the disc or removal or fracture of the abrasive particles will have occurred.

4.7   Number of Tests and Range of Load and or Speed

The scatter is often quite large and it will be necessary to carry out a statistical analysis. However over and above this it is important to evaluate the material(s) over a wide range of conditions. Two materials may appear to be very comparable under some conditions but not under others i.e. high load or speed for example. The methodology outlined in Table 6 should give a good guide.

4.8   Summary on Wear Testing

Whenever possible use one of the devices that have already been documented and on which results and experiences are available, you will then have at least, one other person to compare your work with.

It is very easy to carry out tests without sufficient standardisation of all the parameters involved, i.e. those you do not know about or understand. Under these conditions it is not surprising to find that results are not reproducible.

It is tempting to design a test procedure to produce quick results, but beware that the results obtained are meaningful to the theory or problem that you are investigating. This is particularly appropriate to the effect of frictional heating and repeated stress effects leading to fatigue failure. In this context your best guide is to establish that the service problem is being investigated in such a way that the same failure mechanism is present in both cases.

Measure as many parameters over a wide range of conditions to obtain background experience of wear testing itself before getting into the more difficult area of materials development. (Table 6).

Ensure that your results and conclusions are unambiguous and truly reflect the work carried out, do not draw conclusions which will mislead others.

Friction and wear are not properties of the materials being investigated, they are characteristics of the whole system. Solutions to tribology problems therefore do not necessarily involve improvements in the materials themselves, changes to the operating conditions may be more appropriate and may lead to a more economic solution.

5.   A VERSATILE WEAR TESTING DEVICE

The three most commonly used test methods are (Fig. 6):-

0.1 pin on disc, undirectional

0.2 pin (or block) on ring, undirectional

0.3 pin on plate, reciprocating

One of the most important advantages of these techniques is that they may all be used for fundamental studies, materials development and also simulation studies. These methods can all be used for dry or lubricated studies and may also be used for both adhesion and abrasion evaluation. In all three methods the size and geometry of both pin and its counterface may be varied. The pin may be flat, conical or shaped to the radius of curvature of the ring (pin on ring) particularly where the latter is used for evaluation of bearing materials.

In the case of the evaluation of coatings it is essential to obtain satisfactory alignment between the pin and the counterface otherwise local penetration of the coating may result in its early failure. Background studies of the wear surfaces and debris produced will show how the coating "fails" and if this is a feature of the particular test procedure.

It is advantageous therefore with thin coatings to use a conical rather than a flat faced pin to ensure alignment and reproducibility of the contact conditions. As far as brittle coatings are concerned, damage caused by wear may propagate outside the actual wear track contact area and this may not be recognised or measured by the particular test procedure. For example, this would not necessarily be determined if the wear track width is measured or if a vertical displacement transducer technique is used and would only be revealed by weight loss or by microscopic examination of the surface before and after each test. Weight loss is unlikely to discriminate if the thickness of the coating and the amount of wear is very small in comparison with the total weight of the specimens being tested. These remarks further illustrate that a wide variety of techniques may be required to measure wear occurring on both counterbodies. When wear occurs does it occur by a weight loss or by material displacement or a combination of both.

Consideration must also be given to the effect that any trapped debris might have on subsequent wear behaviour and (Fig. 15) shows how wear is reduced by removing debris from the system. (39) Great care is required therefore in the design of the test as well as design of the counterfaces if the effect of debris generation is to be understood and in this respect conforming surfaces create a greater problem than non-conforming surfaces.

The three methods shown in (Fig. 6) have recently

been incorporated into a single test machine (Fig. 16) which has a distinct advantage in versatility, reduction of capital cost and the elimination of variables from one machine to another. Changes from one mode of test to another can be achieved very quickly through the interchageability of the various test units and by moving the wear arm. Simulation using this equipment is possible when corrosion combines with wear, but is impossible to achieve with the Rubber Wheel test. The latter test can only be used for "low stress abrasion", whilst the equipment shown in Fig. 16 can be used for low and high stress abrasion as well as adhesion and lubricant evaluation.

6. PUBLICATIONS ON TRIBOLOGY TESTING

Apart from the test specifications already referred to earlier there are a number of very useful documents published by ASTM that have arisen out of specific technical meetings usually in conjunction with their committee G6 Erosion and Wear.

SELECTION AND USE OF WEAR TESTS FOR METALS

ASTM STP 615

EROSION: PREVENTION AND USEFUL APPLICATIONS ASTM STP 664

WEAR TESTS FOR PLASTICS   ASTM STP 701

SELECTION AND USE OF WEAR TESTS FOR COATINGS

ASTM STP 769

SLURRY EROSION, USES, APPLICATIONS AND TEST METHODS

ASTM STP 946

SELECTION AND USE OF WEAR TESTS FOR CERAMICS

ASTM STP 1010

At a time when wear testing is becoming more widely discussed these publications are recommended to all those commencing new investigations. The following conclusions may be extracted from these and other sources which help to establish the way forward in test methodology:-

1. RELEVANT AND USEFUL TESTS CAN BE CARRIED OUT BUT
    NOT IN A CASUAL OR UNCONTROLLED WAY.

2. ESTABLISH THE RELEVANCE OF THE TEST TO THE
    APPLICATION.

3. NO UNIQUE TEST EXISTS BUT SEVERAL HAVE SUFFICIENT
    HISTORY SO THAT GUIDELINES HAVE BEEN ESTABLISHED.

The OECD (38) working group on wear has been very
active since the early days of the Jost Report (6) and
nearly two decades ago instituted a round robin wear
test programme. Although the overall correlation from
one laboratory to another was poor, considerable
encouragement was gained from the background information
relating to the reasons for this. It was concluded that
if only one wear mechanism operates the reproducibility
is very reasonable, but utmost care is required to
ensure that temperature and environmental conditions are
adequately reproduced and standardised from one test to
another. In this work there were considerable
differences in the test machines used as well as the
materials investigated which were largely responsible
for the poor correlation. The OCED group is still
active and meets regularly.

7.   A NEW INITIATIVE IN WEAR TEST METHODS

The Versailles Agreement on Advanced Materials And
Standards (VAMAS) is a relatively new international
initiative which has highlighted a number of areas for
special attention one of which is "Wear Test Methods".

The main objectives are:-

1. IMPROVEMENT OF THE REPRODUCIBILITY OF WEAR TESTS BY
    AGREED WEAR TEST METHODOLOGY.

2. CHARACTERISATION OF THE BEHAVIOUR OF 'ADVANCED
    MATERIALS' IN COMPARISON WITH CONVENTIONAL
    MATERIALS. ADVANCED MATERIALS IN THIS CONTEXT ARE
    INORGANIC SURFACE COATINGS AND ENGINEERING
    CERAMICS.

A pin on disc wear test method has been used for

216

many years to investigate tribological characterisations of materials and systems. This test was therefore chosen by the VAMAS "Wear Test" methods working party and a test methodology agreed (Fig. 17). Because the focal point for this work was wear of engineering ceramics, a ball rather than a flat ended pin, on disc was chosen. The pin on disc is one of the most widely used techniques, although most machines operated by different investigators vary in their design and construction and test procedures have not been agreed by prior discussion. The pin on disc method is particularly versatile and may be used for both adhesive and abrasive testing under both dry and lubricated conditions. The same procedure may be used for metals, polymers, ceramics and surface treatments and coatings. It is possible to carry out tests over a very wide range of applied load, speed and environmental conditions and there are a number of well documented examples where the friction and wear tests carried out in the laboratory have been related to particular engineering systems with considerable success.

Friction, wear and temperature may be recorded continuously in the test and transient or transitional effects are therefore easily detected. Wear may also be monitored by weight loss, dimensional and by surface topography changes. Both worn surfaces and the wear debris can be examined after the test to establish the wear mechanism(s) operating during the test. The geometry of the pin can easily be changed by using a flat or a conical end, where the latter may be more appropriate to control alignment between the pin and the disc, or to carry out lubricated tests. Considerable emphasis is placed on methodology and reproducibility by obtaining all materials from a single source, by continuous measurement of friction and wear and examination of wear surfaces and all debris to diagnose the wear mechanism. All seven countries represented in the collaboration have carried out a round robin series of tests to establish repeatability and to lay down the methodology of test and secondly to investigate advanced materials, namely inorganic surface coatings and engineering ceramics. The UK are represented by the author of this paper and the National Physical Laboratory are the co-ordinating body for all UK participants.

The UK is currently represented by five laboratories in the VAMAS programe. The results of a first round robin have been published (39) and the results are extremely encouraging. The Department of Trade and Industry have given financial support for an additional five UK laboratories to up-date their facilities to enable them to participate in the VAMAS programme and when this is completed a total of nine UK laboratories will then be participating. It is under-stood that the USA and the Federal Republic of Germany are exploring the possibility through their relevant committee of establishing the pin on disc as a 'standard' test procedure.

8.   CONCLUSIONS

Tribology tests particularly for wear of materials are currently attracting considerable attention world-wide because of the focus on energy savings, substitution of strategic materials and a desire to introduce advanced materials into industry. In spite of the general scepticism of the value of laboratory wear tests it has been demonstrated that meaningful wear testing can be carried out. It is necessary to continue the trend towards a few standard tests and this includes, particularly for fundamental and materials development, the pin on disc method. Great care is required in producing ranking orders for wear of materials without due reference to the limited conditions of test and their relevance to a particular engineering system. It is necessary to carry out tests over a wide range of conditions and this will ultimately lead to the production of wear maps which can be used as a design aid. It is necessary to use a range of diagnostic techniques to understand and characterise the wear mechanism(s) involved, if the true value of wear testing is to be achieved. The first aim is to establish a test(s) which will form a basis for data to be obtained to enable different materials to be compared one with another. Discussions with suppliers and users currently indicate that we are a long way from achieving this. It is not suggested however that a so called "standard wear test" will provide all the answers for every individual wear situation in industry. This can only be achieved in a step wise process of which the first is the main concern in this paper. It will be necessary for the designer, the supplier of materials and constructors of equipment and finally the user requiring friction and wear resistance to discuss the whole system before an acceptable design is produced. It is however, now possible to provide wear data in the form of "wear maps" for specific combinations of materials and these will provide designers with a base from which they can select and optimise engineering materials for tribological components.

REFERENCES

1. Burnett, P. J., Rickerby, Th. Sold. Flms., 157, (1988), 233-254.

2. Richardson, R.C., Wear, 11, (1968), 245.

3. Lamb, A.D., Mets & Mats., (1987), 138-142.

4. Friction, Wear and Lubrication, Terms and Definitions, OCED, Delft, Netherlands, (1988), 500-509.

5. Ludema, K.C., Lub. Engr.

6. Jost, P., Lubrication (Tribology) (1966) H.M.S.O., London. UK.

7. Noel, R., Ball, A., Wear, 87, (1983), 351-361.

8. Treatise on Materials Science and Technology, Vol. 13, Wear, Wear Resistance of Metals (1979) Academic Press, New York.

9. Source Book on Wear Control Technology, American Society for Metals, Edt. Rigney & Glaeser, (1978), Ohio, USA.

10. Eyre, T.S., Powd. Met., 24, 2, (1981) 57-63.

11. Childs, T. H. C., Trib. Int., 12, (1980), 285.

12. Erickson, R. C., Glaeser, W.A., A.S.M., 8514-004 (1985).

13. Friction and Wear Test Devices, STLE, 838 Busse Highway, Park Ridge, Illinois 60068, USA.

14. Archard, J.F., Research, 5, (1952), 395.

15. Welsh, M.C., Phil. Trans. Roy. Soc. (A), 257 (1965) 31.

16. Eyre, T.S., PhD Thesis, (1971) Brunel University, London, UK.

17. Vichard, J.P., Gaudet, M., Compt. Rend., 260, (1965), 5472.

18. Archard, J.F., Wear, 2, (1958), 445.

19. Archard, J. F., Kirk, M.T., Proc. Roy. Soc. (A) 261, (1961) 532.

20. Sata. T., Wear, 3, (1960), 104.

21. Firkin, E., Wear 6, (1963), 295.

22. Eyre, T.S., Maynard, D., Wear, 18, (1971) 301-310.

23. Eyre, T.S., Baxter, A., Met. Mats., 6, (1972), 435-439.

24. Wright, K.M.R., Proc. Inst. Mech. Engrs. 1B, (1952) 556.

25. Nield, B.J., Griffin, O.G., Wear 4, (1961), 111.

26. Johnson, R.L., Swikert, M.A. and Bisson, E.E. Tech. Rep. Nat. Adv. Comm. Aero., Wasington, D.C., USA. (1947), No 1442.

27. Williams, K., Inst. Mech. Engrs. 2nd Convention, Lubrication & Wear, (1964) Paper No 1.

28. Bowden, F.P., and Freitag, E.H., Proc. Roy. Soc. (A), 248, (1958), 350.

29. Sauer, F.M., Stanton Res. Inst. NAVORD Rep. No. 5452, (1957).

30. Furze, D., PhD.,(1988) Thesis, Brunel University, London, UK.

31. Handbook of Chemistry and Physics, (1962) (44th ed.) p. 2595, Chemical Rubber Publishing Co., Cleveland, Ohio, USA.

32. Reichenbach, G.S. and Pourney, J.L., Lubr, Engr. 20, (1964), 456.

33. Mitchell, L.A., and Crawford, T.S., Inst. Mech. Engrs. 5th Conference, Lubrication and Wear, (1967) Preprint, p.28.

34. Cornelius, D.F., and Roberts, W.H., Trans. Amer. Soc. Lub. Engrs., 4, (1961), 20.

35. Duffrane, K.F., Ceram. Eng. Sci. Proc. 7, pt 7-8, 1052-1059.

36. Matharu, C., PhD Thesis, Brunel University (1989), London UK.

37. Wilson, F., Eyre, T.S., Wear, 14, (1969), 107-117.

38. Synopsis of the results from an International Cooperative Wear Programme, Begelinger, A., et al, Lubr. Eng. 26 (1970), 56-63.

39. Multilaboratory Tribology: Results from the Versailles Advanced Materials and Standards Programme on Wear Test Methods, H. Czichos, et al, Wear 114 (1987) 109-130.

TABLE 1 - HARDNESS OF ABRASIVES AND SECOND PHASES*

Minerals!	Hardness		Material or phase	Hardness	
	Knoop	HV		Knoop	HV
Talc	20		Ferrite	235	70-200
Carbon	35		Pearlite, unalloyed		250-320
Gypsum	40	36	Pearlite, alloyed		300-460
Calcite	130	140	Austenite, 12% Mn	305	170-230
Fluorite	175	190	Austenite, low alloy		250-350
Apatite	335	540	Austenite, high Cr iron		300-600
Glass	455	500	Martensite	500-800	500-1010
Feldspar	550	600-750	Cementite	1025	840-1100
Magnetite	575		Chromium carbide $(F3, Cr)_7C_3$	1735	1200-1600
Orthoclase	620		Molybdenum carbide MO_2C	1800	1500
Flint	820	950	Tungsten carbide WC	1800	2400
Quartz	840	900-1280	Vanadium carbide VC	2660	2800
Topaz	1330	1430	Titanium carbide TiC	2470	3200
Garnet	1360		Boron carbide B_4C	2800	3700
Emery	1400				
Corundum	2020	1800			
Silicon carbide	2585	2600			
Diamond	7575	10,000			

TABLE 2 - COEFFICIENT OF FRICTION AND WEAR RATE OF SOME METAL COUPLES

Metal combination	Coefficient of friction, u	Wear rate, cm^3/cm x 10^{-12}
0.2% carbon steel on itself	0.62	157,000
60% Cu-37%Zn-3%Pb on steel	0.60	24,000
Stellite on carbon steel	0.60	320
Ferritic stainless steel on carbon steel	0.53	270
Tungsten carbide on itself	0.35	2

1.	Clean metals in vacuum	1 to 5 or greater
2.	Metals in oxygen	0.4 to 1.0
3.	Polymer on metal	0.1 to 0.3
4.	Boundary lubrication	0.1
5.	Hydrodynamic lubrication	<0.05

TABLE 3 – TECHNIQUES FOR PRACTICAL WEAR ANALYSIS

Measurement of wear	Appearance of wear	Properties of elements	Description of working conditions
Weighing, linear displacement, wear profiling and isotope techniques	Visual observations, optical microscopy and scanning electron microscopy.	Surface analysis techniques (electron probe microanalysis, energy-dispersive X-ray analysis, Auger electron spectroscopy and X-ray photoelectron spectroscopy) Vickers' hardness adhesion scratch test of coating.	Calculation-measuring of operational variables (load and velocity), materials (lubricants, particles and corrosive media) and vibration-stability.

TABLE 4 - WEAR RATE OF PINS OF VARIOUS MATERALS (AT A LOAD OF 400 GRAMS AND SLIDING SPEED OF 180 CM/SEC)

Combination of Material	Wear Rate $(10^{-10}$ cm^3/cm)	Hardness $(10^6$ gm/cm$^2)$	Calc. Value of K
Mild steel on			
mild steel	1570	18.6	7×10^{-3}
60/40 Brass	240	9.5	6×10^{-4}
Teflon	200	0.5	2.5×10^{-5}
70/30 Brass	100	6.8	1.7×10^{-4}
Perspex	14.5	2.0	7×10^{-6}
Moulded Bakelite			
x 5073	12.0	2.5	7.5×10^{-6}
Silver steel	7.5	32	6×10^{-5}
Beryllium copper	7.1	21	3.1×10^{-5}
Hardened tool steel	6.0	85	1.3×10^{-4}
Stellite grade 1	3.2	69	5.5×10^{-5}
Ferritic stainless			
steel	2.7	25	1.7×10^{-5}
Laminated Bakelite			
292/16	1.8	3.3	1.5×10^{-6}
Moulded Bakelite			
11085/1	1.0	3.0	7.5×10^{-7}
Sintered tungsten carbide on mild			
steel	0.9	18.6	4×10^{-6}
Laminated Bakelite			
547/1	0.4	2.9	3×10^{-7}
Polyethylene	0.3	0.17	1.3×10^{-7}
Sintered tungsten carbide on sintered			
tungsten carbide	0.0	130	1×10^{-6}

225

TABLE 5 - SUBSTANCES GIVING % RELATIVE HUMIDITY AT A GIVEN TEMPERATURE WITHIN A CLOSED SPACE WHEN AN EXCESS OF THE SUBSTANCE INDICATED IS IN CONTACT WITH A SATURATED AQUEOUS SOLUTION OF THE SOLID PHASE

Substance	Temperature ^{0}C	Relative Humidity %
L1Cl. H_2O	20	15
$KC_2H_3O_2$	20	20
$CaCl_2.6H_2O$	20	32.3
"	24.5	31
"	18.5	35
CrO_3	20	35
$Zn(NO_3)_2.6H_2O$	20	42
$NaHSO_4.H_2O$	20	52
$NaBr.2H_2O$	20	58
$NaNO_2$	20	66
$NH_4Cl + KNO_3$	20	72.6
$NaClO_3$	20	75
NH_4Cl	20	79.5

TABLE 6 - VARIABLES IN WEAR TESTING

Load
Velocity
Sliding distance
Temperature
Contact area
Geometry
Surface finish
Atmosphere
Materials
Counterface - 1)
Counterface - 2) Debris
Type of lubricant

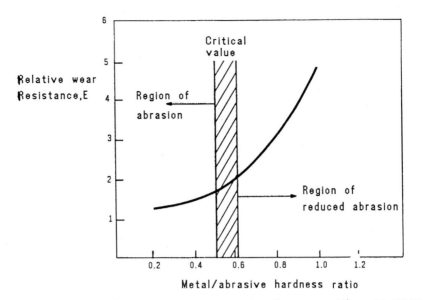

1. Effect of metal/abrasive hardness ratio on wear resistance.

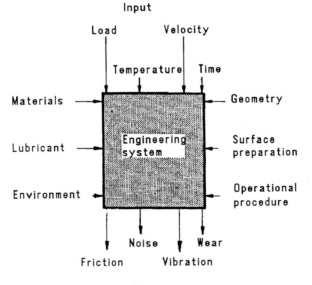

3. Systems approach in tribology.

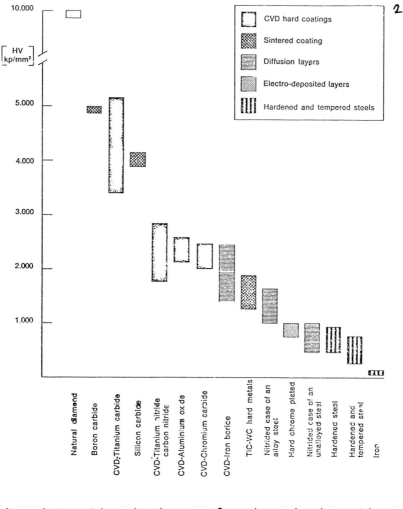

2

Legend:
- CVD hard coatings
- Sintered coating
- Diffusion layers
- Electro-deposited layers
- Hardened and tempered steels

Y-axis: HV kp/mm² (10.000, 5.000, 4.000, 3.000, 2.000, 1.000)

X-axis categories:
Natural diamond, Boron carbide, CVD-Titanium carbide, Silicon carbide, CVD-Titanium nitride carbon nitride, CVD-Aluminium oxide, CVD-Chromium carbide, CVD-Iron boride, TiC-WC hard metals, Nitrided case of an alloy steel, Hard chrome plated, Nitrided case of an unalloyed steel, Hardened steel, Hardened and tempered steel, Iron

2. Comparative hardness of various hard coatings and materials.

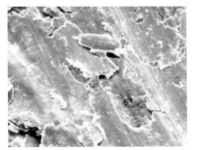

4. Surface appearance of different types of wear (a)
   abrasion (b) adhesion (c) oxidation (d) delamination
   (ℓ) fatigue.

Class	Description	Typical Device & Title	
A	Multiple Sphere		Four Ball Lubricant Tester
B	Crossed Cylinders		Crossed Cylinder Test Apparatus
C1	Pin on Flat (Moving Pin)		Pin Fretting Test
C2	Pin on Flat (Moving Flat)		Kinetic Boundry Friction Apparatus
C3	Pin on Flat (Multiple Contact)		Oxide Scale Friction Apparatus
D	Flat on Flat		Stick - Slip Test Apparatus
E	Rotating Pins on Disc (Face Loaded)		Friction & Cold Welding Apparatus
F	Pin on Rotating Disc (Face Loaded)		Disc Rider Friction Apparatus
G	Cylinder on Cylinder (Face Loaded)		Solid Lubricant Test Machine
H	Pin on Rotating Cylinder (Edge Loaded)		Pin on Ring Tester
I	Flat on Rotating Cylinder (Edge Board)		Dual Rub Shoe Tester
J	Disc on Disc (Edge Loaded)		Rolling Disc Machine
K	Multiple Specimens		Cylinder on Ball Rolling Fatigue Test
L	Miscellaneous		Abrasive Belt Wear Test Machine

5.   STLE classification of wear devices.

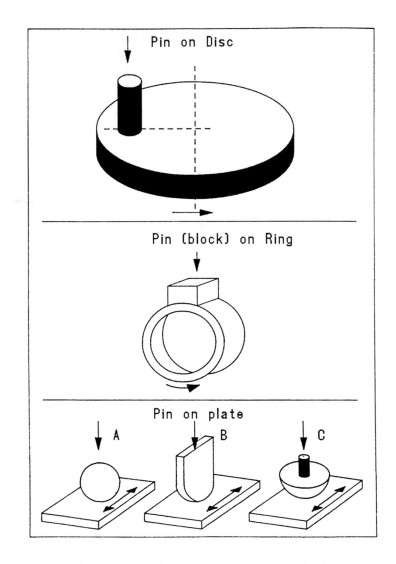

6. The three most widely used wear devices.

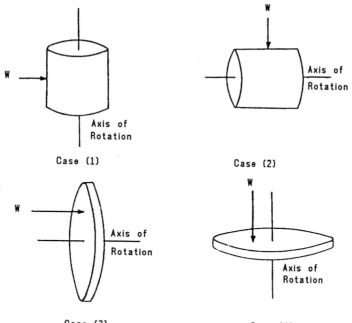

7. Load application configuration in tribo testing.

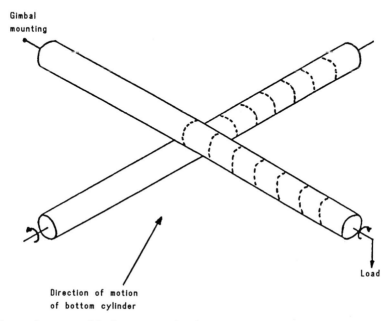

8. Cross cylinder wear device.

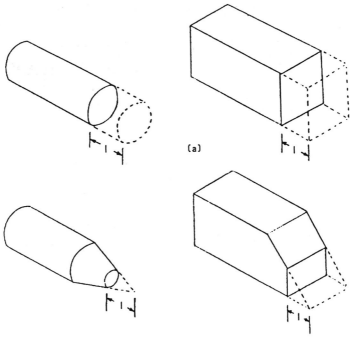

(a)

(b)

9.   Different wear pin configurations.

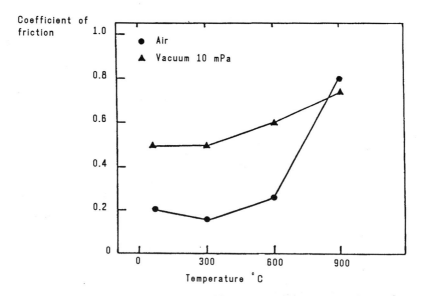

Coefficient of friction

● Air
▲ Vacuum 10 mPa

Temperature °C

11. Variation of friction coefficient with temperature for RBSN (reaction bonded silicon nitride).

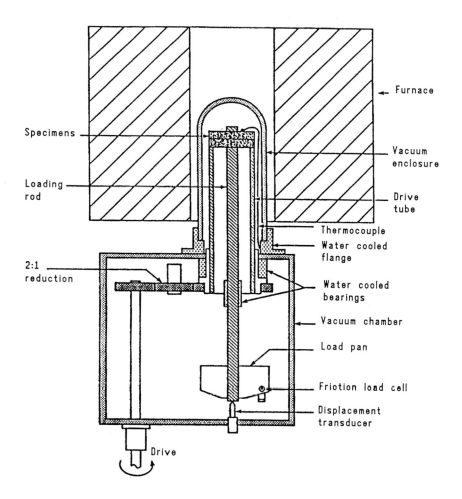

Furnace

Specimens

Vacuum
enclosure

Loading
rod

Drive
tube

Thermocouple

Water cooled
flange

2:1
reduction

Water cooled
bearings

Vacuum chamber

Load pan

Friction load cell

Displacement
transducer

Drive

10.   High temperature vacuum tribo tester.

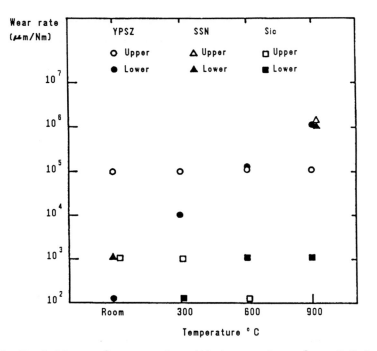

12. Variation of wear rate with temperature for Y.P.S.Z. (Yttria partially stabilised zirconia), SSN (sintered silicon nitride) and SiC (silicon carbide).

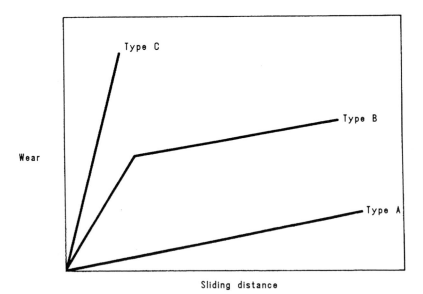

13. Schematic of the three most widely encountered wear versus sliding distance relationships for metals.

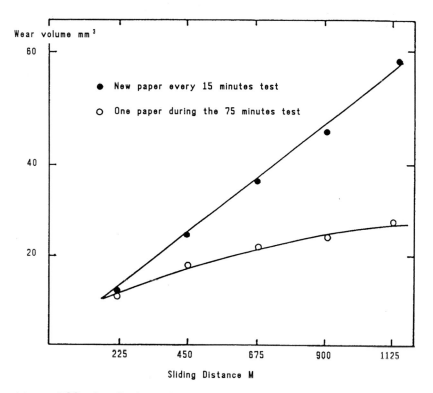

14. Effect of changing the abrasive cloth in abrasive wear testing of steel.

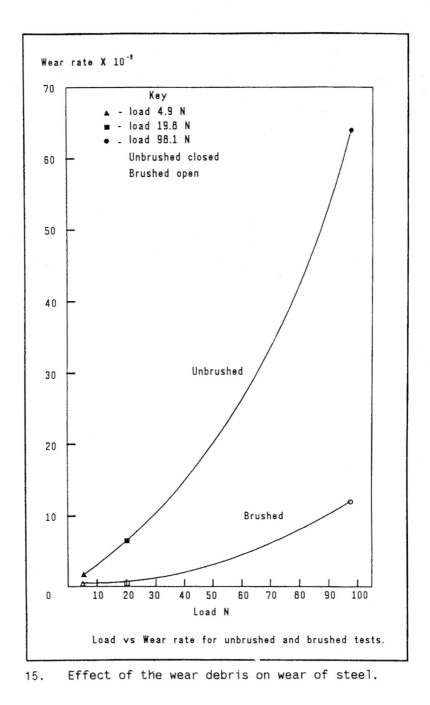

Load vs Wear rate for unbrushed and brushed tests.

15.   Effect of the wear debris on wear of steel.

16.   Tribo-testing machine in the pin on disc mode.

VAMAS project "Wear test methods" DATA SHEET			Comments
Test system	F_N ←— 32 mm —→ v		
Geometry	ball (stationary)	On disc (moving)	
Dimensions	Ø 10 mm (or Ø 1/2")	Ø 40 mm 6 mm thick	
Materials	AISI 52100 steel	steel SiC TiC on steel Others:	
Surr.atmosphere	Laboratory air (50 ±10% rel.humidity T = 23 ± 1°C)		
Lubricant	None	Mineral oil (SAE 10)	
Operating variables			
Motion	Cont. sliding		
Velocity v	0.1 m/s		
Normal Load F_N	10N	300 N	
Sliding distance	1 km		
Number of tests	5	3	
Mearurements	Friction force (continuous) wear of both partners (vol. and weight loss) wear surfaces (SEM, profilograms)		
Institution:			
Signature:		Date:	

17.   VAMAS data sheet.

6:Surface analysis of advanced materials and their degradation products
R K WILD

1.   INTRODUCTION

The last twenty years has seen a revolution occur in the use of new materials. This has occurred in all areas of life where materials that were not thought of in the fifties are now commonplace. In the field of transportation vast changes are taking place. Aircraft are constructed from novel alloys such as titanium, aluminium and an increasing fraction of the body is constructed of plastics and composites. Indeed it has been predicted that by the year 1995 50% of an aircrafts weight will come from aluminium alloys, 30% from composite materials, 10% from titanium alloys and only 10% from what were regarded as conventional materials.   Cars are also increasing the use of plastics and are close to developing ceramic engines and composite gearboxes and even the humble bicycle can be obtained with carbon fibre frames and light alloy components. Energy production has been revolutionised with steadily increasing efficiency. This has meant in the electricity industry power stations operating at higher temperatures which in turn requires new alloys to withstand the corrosive environments. Here increasing use is being made of the new nickel based superalloys. The oil industry has had to search for oil in increasingly hostile environments which has again meant that the old materials cannot be used. However, new materials are not a panacea for all the problems encountered with the old alloys. They themselves may be subject to degredation and stress corrosion cracking has proved to be a problem with some nickel based alloys. Many of the new composite materials corrode at unacceptably high rates. These difficulties may be overcome by understanding the microstructure of the material and how it changes with time and temperature in the operating environment. The cause of the problem may be a trace element segregating to an interface and removal or restriction may be possible. To understand the microstructure requires the use of new analytical techniques which have been developed alongside the new materials. Techniques are required that can identify all elements, provide chemical state information, can detect this when the element is present in small quantities and in single atom layers and do all this on both external and internal interfaces. In the last two decades techniques have been developed which provide most of this information although there is still some development required. This paper is intended to describe how advanced materials may be characterised using surface analytical techniques, how the degredation products produced during their use may be identified and how remedial measures can be taken. Where possible this will be illustrated with examples.

2.   ULTRA HIGH VACUUM AND CLEAN SURFACES

The study of these materials necessitates the production of clean surfaces and maintaining the surface in a clean condition for a period that is sufficient to allow the analysis to be performed. At a pressure of 10^{-5} Nm^{-2} sufficient gas atoms imping on a surface in one second to completely cover that surface if each atom that hits the surface sticks to it. Thus if a clean surface is to remain clean, that is contamination is kept to less than 10% of a monolayer, for say one hour then the vacuum in which the surface is kept must be better than 10^{-9} Nm^{-2}. The production and maintenence of such vacuums required the development of a whole new technology. A typical surface analytical system is shown in Figure 1. The system is constructed almost entirely from nonmagnetic stainless steel and the seal between components is made by knife edges cut into the steel biting into an annealed copper flange. Any translated movement must be through welded bellows and electrical connections by ceramic feedthroughs. Pumping is now generally achieved in three stages. An initial roughing pressure is normally obtained by using a turbo molecular pump which is essentially a high speed turbine which deflects molecules out of the chamber with the blades. In some cases sorption pumps cooled to liquid nitrogen temperatures may be employed. These utilise the large surface area of a molecular sieve to adsorb gases at the low temperature and later desorb them by heating. The next stage, which takes the pressure from 10^{-2} Nm^{-2} down to 10^{-6} Nm^{-2} generally employs an ion pump which ionises the gas atoms and then utilises a high magnetic field to deflect and embedd the ions on the vessel walls. However, conventional oil diffusion pumps using low vapour pressure oils and combined with cold traps may also be used. The final pressure of 10^{-9} Nm^{-2} is reached by baking the entire system to 500K to desorb water vapour from the walls of the system and by subliming titanium into a chamber to speed the removal of active gases.

2.1   Producing Clean External Surfaces

When a sample is introduced to the system it will have many atom layers of contaminant on it. This may be in the form of an oxide but may also contain carbon and water vapour. This can be removed by bombarding the surface with an energetic beam of inert gas ions often produced using an ion gun that may be part of a SIMS unit (see below). Generally argon is the gas ion chosen for the bombardment but specific requirements may utilise helium, xenon etc. Other sample specific methods may be utilised, such as heating in hydrogen and oxygen. For example, sulphur can be removed from the bulk by heating in oxygen but this forms a surface oxide which can in turn be removed by heating in hydrogen. However, such methods tend to significantly modify the matrix.

2.2   Producing Internal Clean Surfaces

Two types of internal surface can be envisaged. One is the interface between two dissimilar layers such as that between an oxide and metal or between a fibre and a matrix while the other is the interface between similar layers such as grain boundaries in alloys. Exposure of each interface requires its own specific method. We have made use of a technique known as Sputter Ion Plating (SIP) (see below), to expose the interface between an alloy matrix and its oxide. Here, (Figure 2), a metal layer is deposited onto the outer surface of the oxide at a high temperature and in the presence of an argon glow discharge. The glow discharge cleans the surface and at the same time produces a strong bond between the metal and oxide. On cooling to room temperature the stresses are set up in the layers that allow the deposited metal to be used to peel the oxide from the matrix. Figure 3 shows a secondary electron image of the two sides of the metal/oxide interface on a high nickel stainless steel.

However the most frequently studied internal interface is the grain boundary. Again specific methods must be used to expose these. For example a grain boundary in a ferritic steel may be exposed by impact fracture at liquid nitrogen temperature but the equivalent boundary in a nickel based superalloy can only be exposed by a slow tensile strain following hydrogen embrittlement (see below). Often the interface between oxides and alloys can be exposed by a technique known as sputter ion plating while the interface between a fibre composite and its matrix may be accessible following either an impact or tensile fracture.

It cannot be stressed too often that the interface must be exposed in the ultra high vacuum environment. To demonstrate this I reproduce a figure from an experiment performed by Seah [1] (Figure 4). He produced a clean iron surface in the UHV environment onto which was deposited a monolayer of tin. He obtained the Auger spectrum (a) from this surface and the tin peaks are clearly visible at 430 and 437 eV. This surface was then exposed to 160kPa of air at room temperature and the Auger spectrum recorded again. The resultant spectrum (b) shows that the tin peaks have decreased to less than 5% of the original intensity. Tin is one of the classical embrittling elements and had a fracture surface been exposed outside the vacuum no tin would have been detected. It is therefore essential that all fracture surfaces be exposed at pressures $< 10-8$ Pa.

3.   TECHNIQUES

There is a whole host of techniques available for application to surfaces and interfaces. However some have gained greater acceptance than others in recent years. In the early days of surface analysis Xray photoelectron spectroscopy (XPS), Low energy electron diffraction (LEED), Field ion microscopy (FIM) and Auger electron spectroscopy (AES) were the major surface analytical techniques. Today LEED and FIM are less frequently used but secondary ion mass spectrometry (SIMS) and the field emmission gun scanning transmission electron microscope (FEGSTEM) have joined the arsenal while a time of flight mass spectrometer has been added to the FIM to produce an atom probe. Space does not permit a full description of these techniques and is only sufficient for a brief description of the principals and operation of the three techniques most frequently used today namely XPS, AES and SIMS.

4.   X-RAY PHOTOELECTRON SPECTROSCOPY (XPS).

The principle of this technique [2] is shown schematically in Figure 5. A beam of monoenergetic photons of energy hv is focussed on the surface. This causes atoms in the surface layers to be ionised by the ejection of either inner shell or valence electrons. The ejected electron has an energy given by:

$$E = hv - E_b - \phi \tag{1}$$

where ϕ = the surface work function.

E_b = Binding energy of ejected electron.

Thus by knowing the incident photon energy and measuring the ejected electron energy the electron binding energy can be deduced and the surface atom identified. In addition, changes occur in the electron binding energies when two dissimilar atoms combine. By detecting these changes in binding energy the chemical state of the surface atom can be determined. Figure 6 shows schematically the layout of a modern X-ray photoelectron spectrometer. In this technique the sample lies on a manipulator, usually in the horizontal plane, and is bombarded by X-rays. The ejected electrons are generally focussed onto the input lens of a hemispherical analyser. This focussing lens also retards the electrons so that the analyser can be operated at a fixed hemisphere voltage. In this way a constant energy resolution is obtained over the entire spectrum. The electrons which pass through the analyser are detected using an electron multiplier arrangement. With a hemispherical electrostatic analyser of this type the electron binding can be determined to within $\mp$ 0.1

eV which is sufficient to detect most chemical state changes in solids. To improve the energy resolution further many instruments include an X-ray monochromator. The incident photon is normally the Kα radiation from either magnesium or aluminium and has a kinetic energy of 1253.6 ev or 1486.6 ev respectively. Thus all electrons ejected from the surface have a kinetic energy of < 1.4 keV. Such electrons have a mean free path of less than 1.5 nanometers and hence the signal detected emanates from the top few monolayers.

To illustrate the type of results that are obtained from an XPS system a series of spectra are shown in Figure 7 which were taken from the surface of a high nickel chromium steel while the clean surface was exposed to oxygen[3]. The top spectrum is a wide scan survey of the clean surface, which would normally be used to identify which elements were present, and shows photoelectron peaks from the nickel, iron and chromium together with Auger peaks from iron (see below) at 600, 650 and 700 eV kinetic energy. Auger peaks can be identified from photoelectron peaks by changing the energy of the incident photoelectron, by switching from say magnesium to aluminium radiation. When this is done the photoelectron peaks move an amount equal to the change in the photoelectron energy but the Auger peaks remain at a fixed energy. As the surface is exposed to oxygen so the oxygen 1s photoelectron peak and oxygen KLL Auger peaks increase in intensity and the nickel and chromium peaks decrease. This indicates that the oxide forming on the outer surface is essentially iron oxide. The power of the technique lies in its ability to detect changes in the peak position when the atom changes its chemical state. This is illustrated in Figure 8 and shows the changes that occur in the oxygen 1s region during the early stages of oxidation of nickel. Initially the oxygen forms a non dissociated adsorbed layer of oxygen then oxygen begins to dissociate and form chemical bonds and the oxygen 1s peak moves to a binding energy 1.7 eV lower. At room temperature a layer of oxide forms with most oxygen atoms on the outer surface. A true nickel oxide is formed by heating to 550K when diffusion occurs and this is reflected in the 1s peak which shifts back to a higher binding energy[3].

One problem with this technique has been the lack of spatial information. It is not possible to focus the incident photoelectron beam and until recently it has only been possible to obtain spectra from the entire area illuminated by the incident beam. However by collimating and making use of specific electron lenses in front of the hemispherical analyser the area analysed has been dramatically reduced. In recent years this area has been continuously reduced, first to 1mm x 1mm and then 100 μ m x 100 μ m and in the last few months an instrument with less than 10 micron spatial resolution has been announced.

4.1 Auger Electron Spectroscopy (AES).

In this technique[2] a beam of electrons is used to ionise the surface atoms but the electron directly ejected is not used to identify the atom. Instead the Auger electron is detected. Since the energy of the Auger electron depends only on the electron binding energies within the atom it is not necessary to know the energy of the incident electron beam. One simply has to have suffient energy to efficiently ionise surface atoms. The Auger process is shown schematically in Figure 9. Following ionisation by the incident electron the atom relaxes with an electron from a higher electron shell (L shell) falling into the hole in the inner shell (K shell) and in so doing releases energy which is transferred to an outer shell electron (M shell) which may be ejected with an energy given by:

$$E_{KLM} = E_K - E_L - E_M - \phi \qquad (2)$$

where E_K, E_L, E_M are binding energies of the
K,L,M electron shells.
ϕ is the work function.

Figure 10 shows schematically the construction of an Auger
spectrometer. In this instrument samples are introduced to the
analytical chamber through an introduction chamber. The sample is
transferred to a manipulator which may take it to various stations
for preparation. In this case a tensile fracture stage is illustrated. The
sample, when ready, is moved to the Auger analyser. The electron
beam is incident normally on the surface and the secondary electrons
are analysed using a cylindrical mirror analyser (CMA). This
analyser is concentric with the electron gun and electrons enter the
entrance slit at the front of the inner cylinder are deflected to pass
through the exit slit where they are detected with an electron multiplier. This type of analyser has excellent transmission but poor
resolution. In recent years hemispherical analysers have tended to
replace the CMA because of their considerably better energy resolution and recently improved detection efficiency.

In the early days of Auger spectroscopy it was customary to
differentiate the collected signal. This was because the Auger peaks
were superimposed on a large slowly varying background. Differ-
entiation was carried out electronically using phase sensitive detectors. Today the signal is collected in the undifferentiated form but
it is still common to present the spectra in the differentiated mode
using the computer data system. Figure 11 shows a spectrum from
a titanium carbide on a fracture surface in both the undifferentiated
mode (N(E)) and the differentiated mode (dN(E)/dE). As with the
XPS technique electrons with energies between 0 - 2 keV are

detected, ensuring that the depth of analysis is restricted to the top few atom layers.

Again there is the possiblity of detecting changes in the chemical state of surface atoms but to date this has only been attempted in a few cases. There are two reasons for this; firstly the energy resolution has in the past been insufficient and secondly interpretation has more difficulties than with XPS.

All elements with the exception of hydrogen and helium may be detected to a limit of 0.1% of an atom layer. In addition the incident electron beam can be focussed to less than 10 nanometers to give a high resolution secondary electron image and with realistic Auger analysis from areas less than 50 nanometers in diameter. This makes the technique ideal for the study of segregation to grain boundaries in metals and alloys and for analysis of the interface between composites and their matrix.

Quantification is relatively straightforward. There is only a small variation in the sensitivity of the technique for element detection throughout the periodic table and matrix effects are small. Considerable effort has been devoted to obtaining accurate sensitivity factors with the result that in uniform layers accuracies of $\mp$ a few percent are routine.

4.2   Scanning secondary ion mass spectroscopy (SSIMS).

In this technique[4], shown schematically in Figure 12, a beam of energetic ions with energies of up to a few keV is directed onto the surface. This causes ions and ion clusters to be ejected from the surface. These are then detected and analysed for mass. The mass analysis can be achieved in several ways. In the simplest a quadrupole mass spectrometer is employed often with a 90° prefilter.

In instruments with higher mass resolution a magnetic sector mass spectrometer is employed. However, the most sophisticated SIMS instruments utilise a time of flight (TOF) mass spectrometer. The TOF can detect and identify individual ions and gives the technique excellent sensitivity. Clearly this technique has the drawback that it is destructive but removal rates can be sufficiently slow, particularly when the TOF is used, that only fractions of a monolayer are removed. The technique is also highly sensitive being capable of detecting part per billion levels for certain elements although sensitivity variations between elements makes quantification a nightmare. There is a variation in sensitivity of detection between an element with poor sensitivity and an element with good sensitivity of up to four orders of magnitude. In addition there are dramatic changes in sensitivity when comparing the same element in different chemical states. However, by studying the "finger print" spectrum from a surface chemical state information can be obtained with some degree of quantification. This does, however, require the production of known standards but is a well established method.

Normally the incident ion beam is produced from an inert gas atom such as argon or helium. In such systems the ion beam may be focussed to a diameter of a few microns which in turn may be rastered to give an image from the surface, however the spatial resolution is not quite good enough for the study of internal surfaces. In recent years liquid metal ion sources have become available using gallium, indium, gold, etc. . These can be focussed to less than 0.1 micron and produce good images with excellent contrast and depth of focus. This is illustrated in Figure 13 which show a SIMS image taken with a gallium source liquid metal gun together with the equivalent SIMS spectrum from the surface of one of the new high temperature superconductors. This particular superconductor has a bulk composition $Bi_2-xPb_xSr_2Ca_2Cu_3O_y$ The good spatial resolution can be seen in the SIMS image, recorded by detecting the secondary electrons produced while rastering the gallium ion beam. The spectrum contains peaks from all the major elements; Ca(40,42,44), Cu(63,65), Sr(84,86,87,88), O(16) and not shown Pb(208) and Bi(209). In addition peaks are detected from ion clusters CaO(56), CaOH(57), $Ca_2O(96)$, SrO(104), SrOH(105), $Ca_2O_2(112)$, CuCaO(119,121), GaCaO(125,127), and SrCaO(144). The peak at mass number 23 arises from sodium. This element has one of the highest sensitivity factors for any element and the small peak height indicates that this compound is very clean. This technique has many interpretational problems but offers the possibility of the most sensitive probe for segregating species to grain boundary surfaces.

5.   EXPOSED SURFACES

The outer surface of a material is clearly the simplest to analyse and detect. The composition and formation of oxides may be determined either in situ or by oxidising in a special cell and the moving the sample for analysis. The composition with depth can be determined by profiling down through the surface using an ion beam. An example of the study of the oxidation of a high chromium stainless steel used in corrosive gaseous environments at temperatures up to 1100K is given in Figure 14 [5]. The surface is relatively flat but following oxidation it contained a large number of small particles as shown in the secondary electron image. XPS analysis shows the surface to form an oxide rich in chromium and manganese which can be identified as a spinel of the form $MnCr_2O_4$ but this did not identify the surface particles. However, scanning Auger microscopy (SAM) was able to focus on individual small particles in the surface which were identified as niobium oxy-carbides from the Auger spectrum.

Segregation to outer surfaces can be readily detected using these techniques [6]. Figure 15 is an Auger spectrum recorded from the outer surface of stainless steel following heating in vacuum to 973K. It can be seen that sulphur, with an Auger LMM peak at 150 eV, has covered the surface. In this experiment a cold finger of aluminium was placed close to the heated steel and was analysed following heating of the stainless steel. In the spectrum (Figure 15b) recorded from the cold finger large LMM Auger peaks are observed from manganese at 542, 589 and 636 eV and from chlorine at 179 eV are detected. These elements are effusing from the stainless steel during annealing. Manganese is an austenite stabilising element and prolonged annealing in vacuum can cause the structure of the stainless steel to change from face centred cubic to body centred cubic. This is a good example of the way in which surface analysis can be used to predict bulk metallurgical problems and indicates that care must be exercised in the type and form of the heat treatments given.

The effect of sulphur segregation on the early stages of oxidation can be seen by monitoring the surface composition while exposing to a low oxygen partial pressure [6](Figure 16). Sulphur is initially present in large quantities and covers the surface to a depth of only one atom layer but to a concentration greater than half a monolayer. When oxygen impinges on the surface it reacts with the sulphur to form SO_2 which is released as the gas. Sulphur, at the surface, is continuously replenished from the bulk until the underlying matrix is depleted of sulphur. Only at this time is the oxygen able to form an oxide with the steel constituents. At first the most thermodynamically stable oxide, Cr_2O_3, forms and indeed the ratio of oxygen to chromium after 300 minutes is 1.33, but gradually manganese becomes incorporated into the oxide and the spinel $MnCr_2O_4$ forms. This is a good example of the quantitative nature of Auger spectroscopy. The oxygen content of the oxide drops from 60% in the Cr_2O_3 to 57% in the spinel and this change is readily detected by the technique.

6.   HIDDEN SURFACES

Hidden or buried surfaces present particular problems. First there is the buried interface between two disimilar layers. This may be the interface in a semiconductor between layers with different dopants, or it may be the interface between an oxide and the underlying matrix or the interface between a fibre and the matrix. The interface can be reached by profiling down using an argon ion beam [7] but if the interface is more than 1 μm below the surface mixing of the materials reduces the resolution and segregants may be missed. However, this technique is the most commonly used to detect and identify the changes that occur between different layers. An example of a profile down through a material that contained many layers closely spaced is shown in Figure 17. This is a profile through 40 layers on a silicon substrate. The layers are alternately silicon and titanium-zirconium. Each layer is only 25 atom layers thick and it can be seen that these layers have been easily resolved down to the silicon substrate, a depth of 300 nanometers below the surface. Towards the end of the profile there is some evidence of mixing of the layers. Other related techniques for detecting buried surfaces include the initial production of a taper section through the layer which is then cleaned by ion etching and the layer studied using the high spatial resolution of an Auger microprobe. The taper section can be either a straightforward metallurgical polish at a narrow angle or a rotating hard sphere may be used to produce a shallow crater (known as ball cratering).

An alternative method of reaching the interface is to pull the layers apart. This can be done if the interface is a layer of weakness but would be unlikely to be successful if the layers are forming strong bonds with the matrix on either side. In our own work we have employed the technique of sputter ion plating (SIP), see below, to coat an oxide with nickel. This is done at 500K in a plasma discharge of argon and gives a good bond between the nickel and the oxide layer. On cooling, stresses set up between the metal and oxide layers, cause the oxide/nickel layer to pull apart revealing the interface. An example is given earlier in Figure 3. Here, a 20%Cr/25%Ni/Nb stabilised steel has been oxidised at 823K for 12000 hours. The oxide has pulled away from the metal to reveal the underlying grain structure. Auger spectroscopy has revealed that at the interface between the oxide and the metal a very thin layer of silicon oxide exists. This layer is only 10-30 nanometres thick but is the line of weakness. When the oxide pulls apart it does so by fracturing through the thin silicon layer on grain surfaces but at grain boundaries the forces are such that failure occurs through the overlaying chromium oxide. This is good news for cases where the oxide spalls off because the chromium oxide "plugs" in the grain boundaries reduce grain boundary cation diffusion and hence reduce subsequent oxidation rates.

The second type of hidden surface is the grain boundary. It is also the most important interface in the study of metals and alloys since it determines the mechanical properties. Fracture is a popular method for exposing internal grain boundary surfaces. The most commonly used method of fracture is to cool to liquid nitrogen temperature and impact a notched sample. Provided the temperature of the sample is below the brittle/ductile transition temperature the sample will fail with a high degree of intergranular surface. A typical example of the fracture surface is shown in Figure 18a which was taken from a 2.25%Cr/1%Mo steel bolt that had been in service for 50,000 hours. The grains can be clearly identified by their flat surfaces with the presence of carbides being detected as a slight roughening. Figure 18b is an Auger spectrum taken from one of the grain surfaces. Two segregating impurities can be seen, one is carbon which has formed a chromium-molybdenum carbide at the boundary while the other is phosphorus. Phosphorus is a known embrittleing agent and is frequently observered on the grain boundaries of materials which exhibit brittle failure.

Unfortunately intergranular failure does not always occur in alloys by impact at low temperatures. This is particularly true of the new superalloys. These alloys are in general ductile at liquid nitrogen temperatures. It is necessary to hydrogen charge the material before fracturing by a slow tensile strain. This has been achieved in our laboratory by modifying the impact fracture stage in such a way that the sample can be loaded and fractured in tension in less than one hour following hydrogen embrittlement. Clearly there are heat treatments that can cause the material to be brittle and avoid the need for hydrogen charging but these are the exception rather than the rule. Examples of the fracture of a high nickel alloy, PE16, are shown in Figure 19. Figure 19a shows the intergranular surface from a good unaged sample. The grains are smooth and regular and have a composition similar to the bulk. The second sample (Figure 19b) has been aged for several thousand hours at 873K and has a rough surface covered in particles and shows a high level of phosphorus. The particles are chromium molybdenum carbides. Occasionally samples fracture intergranularly without the need to hydrogen charge. This is often the result of heat treatments which have allowed chromium/titanium carbides to form over large areas of the grain boundaries and failure occurres at the interface between these boundaries and the matrix.

Inconel 600 is another nickel based alloy which is used in the construction of boiler tubes for Pressurised Water Reactors. It has been shown to have a tendency to stress corrosion cracking. The cause of the problem was thought to be trace element segregation to grain boundaries allowing enhanced corrosion to take place down a boundary in the presence of stress. To identify which element was responsible a base alloy was produced to which trace quantities of boron, sulphur, phosphorus, silicon and carbon were added. These were then given a solution annealing treatment to 1373K and some of these an ageing treatment at 977K for 20 hours. It was found that boron, phosphorus and sulphur all segregated to the boundaries following annealing and ageing. However, following annealing alone the boron segregated to the grain boundary as free boron but during ageing the boron formed a borocarbide with the added carbon. This can be compared with other observations which show enhanced stress corrosion cracking (SCC) on the annealed material but less SCC on the heat treated samples. It is most probable that free boron is playing an important role but that tieing it up in the form of a carbide reduces the deleterious effect of this element.

7.   CORROSION AND CORROSION PRODUCTS

Many of the new materials have been developed for their improved corrosion properties. This is particularly true of the nickel based alloys which are used in highly corrosive environments that are as diverse as cold salt water to high temperature gases. In addition aluminium and titanium alloys offer excellent corrosion resistance at low temperatures as a result of the healing oxide layer which rapidly forms. However, increasing use is being made of the composite materials which use fibres in aluminium alloy matrices. The corrosion characteristics for many of these composites are very poor and considerable effort will be needed to understand the corrosion processes here. It is interesting to analyse how these alloys corrode, what influences the corrosion and how they may be further improved.

7.1   Titanium/Aluminium Alloys

Because titanium is light, relatively strong and offers good corrosion resistance at high temperatures it is finding increasing use, particularly in the aircraft industry but also in other areas such as body implants. Alloying titanium with aluminium has the effect of increasing the yield strength, the ultimate tensile strength and it also shows improved wear resistance.

Both XPS and AES have been used to characterise the oxide which forms on the alloy Ti6Al4V (8). Figure 20 shows a wide and narrow scan XPS spectrum from the surface oxide after giving the surface a sputter etch to remove 1 nanometer of oxide. This shows major peaks from titanium, oxygen and nitrogen with a smaller peak from aluminium. Narrow scans taken from the titanium 2p, aluminium 2p and nitrogen 1s regions indicate that the oxide is essentially TiO_2 but that aluminium is also present in the form of an oxide and that there is a significant level of nitride, presumably present as TiN. This alloy has approximately 4 wt.% of vanadium in addition to the 6.5 wt.% of aluminium but the vanadium does not appear in the outer oxide layers. The study reported here is on an alloy for use in biological implants and compares the merits of this alloy with pure titanium. Although the new alloy has improved mechanical strength and better wear properties there is the worry that the aluminium in the outer oxide may dissolve into the surrounding biological medium, with consequent health problems.

7.2   Nickel Based Alloys

When nickel or nickel containing alloys are heated to temperatures in excess of 400°C sulphur tends to segregate to the surface. The level of sulphur segregation in absolute terms has been determined by Seah [9] and compared with the sulphur segregation to grain boundaries. This is reproduced in Figure 28. It was shown earlier that sulphur segregation can have an influence on the oxidation mechanism when the alloy is exposed to a corrosive environment. The kinetics of oxidation can be studied using surface analytical techniques. High nickel alloys containing iron and chromium normally form an iron rich spinel oxide layer which grows relatively rapidly. It is only when the chromium forms a healing layer below this spinel that the oxidation rate decreases. A first step towards understanding the oxidation process is to characterise the oxides. The oxide which forms on such alloys is very thin, less than 2μ m thick, and it is not possible to use conventional metallurgical techniques to characterise the oxide. However, by using a depth profile with either XPS or AES analysis and in combination with X-ray diffraction the composition and structure as a function of depth may be determined [10]. Figure 21 shows an element depth profile on such a steel together with X-ray diffraction patterns obtained before depth profiling and after profiling to a depth of 0.4 μm. The first diffraction pattern shows peaks from the spinel oxide, the rhombohedral oxide and the underlying austenite. After removal of 0.4 μm the spinel peaks have been dramatically reduced in intensity indicating that the outer oxide layer is the spinel and hence the inner layer must be the rhombohedral.

7.3 Surface Treatments to Reduce Corrosion.

7.3.1 Preoxidation Treatments

By giving the alloy a preoxidation treatment it is possible to form a protective oxide coating which will reduce subsequent oxidation rates. This has been done for the alloy 20%Cr/25%Ni/Nb steel both in our laboratory and elsewhere [10,11]. In each case a protective chromium rich layer was first put down by heating the steel in a low oxygen partial pressure. The steels were then exposed at high temperature to the high pressure corrosive environment and the weight gain compared with material that had not undergone the pretreatment. It can be seen from Figure 22 that the corrosion resistance has been improved by a factor of 3 to 4.

7.3.2   Ion Implantation.

Implantation of certain ions into the steel surface can also dramatically improve the corrosion characteristics of the high nickel, chromium steels [12]. In particular cerium, yttrium and lanthanum can have a beneficial effect on the corrosion resistance. These ions, when implanted into the surface become incorporated into the oxide scale near to the metal/oxide interface and also segregate to grain boundaries. The implanted reactive elements extend the protective nature of the oxide to 1000°C by modifying the initial scale growth and resultant scale microstructure. SIMS has been used to study the effects of implantation of these ions. Figure 23 is a dynamic SIMS depth profile through the oxide scale formed on 20%Cr/25%Ni/Nb stainless steel which had been implanted with 10^{17} yttrium ions per cm^2 and then oxidised for 216 hours in carbon dioxide at 900°C. yttrium can be seen to be incorporated in the outer spinel oxide and to a lower concentration in the inner rhombohedral oxide. The yttrium concentration rapidly drops to zero concentration in the bulk matrix. Both the reactive elements, lanthanum and cerium, act in a similar manner to reduce corrosion.

7.3.3 Surface Coatings

While ion implantation is one method of improving the corrosion characteristics of materials it is often possible to obtain beneficial results by coating the surface. Coatings by plasma vapour deposition (PVD) of silica layers have offered improvements but mechanical damage to the coatings often resulted in localised attack. Coatings which include the reactive elements cerium, yttrium and lanthanum can also prove beneficial. In particular sol-gel coatings have been tried at Harwell Laboratories[13]. In one example shown here sol-gel coating of CeO_2 onto nickel inhibited scale growth by nucleation of cerium oxide particles, the blocking of short-circuit diffusion paths by segregation of cerium ions and by reducing the stresses in scales through changes in growth processes and microstructure. Figure 24 shows a SIMS composition depth profile for the oxide scale formed at 900°C for 142 hours on a CeO_2 sol-gel coated nickel sample. There is a maximum in the Ce distribution at the position of the original CeO_2 layer but diffusion of Ce has clearly occurred both into the inner NiO layer and to the outer oxide.

8.   WEAR PROPERTIES

The performance of a material is the result of a number of properties including strength, toughness and wear. The latter is of prime importance when the material is used to construct moving parts. Here the surface is of importance and surface science can aid the materials engineer. In particular the effects of wear can be identified by using surface analytical techniques to identify the component parts of a wear scar. This aids the interpretation of the wear process and may lead to improvements in the surface treatments to reduce wear.

8.1 Ion Implantation

One technique for improving the mechanical properties of a surface is ion implantation. Many ions have been experimented with and significant improvements in the wear properties have been achieved. Probably the most commonly used ion is the nitrogen ion. This is readily available, easy to implant and will combine with surface atoms to produce a strong outer layer. The ion implantation using nitrogen is particularly successful with titanium and titanium alloys. Here the nitrogen combines with the titanium metal to form an extremely hard layer of titanium nitride that is a few micro metres deep. This has been successfully applied to surgical implants where the life limiting factor is the wear rate. This is particularly the case with hip joint replacements.

The hardness of the titanium nitride layer has lead to a further developement which combines both ion implantation and surface atom coating and is known as sputter ion plating (SIP). Sputter ion plating takes place in a vacuum chamber containing argon at a pressure of 2 Pa. A glow discharge is then initiated with titanium plates acting as the cathodes. This ionises the argon ions which bombard and sputter the titanium atoms. At the same time nitrogen is bled into the chamber. This combines with the titanium to form titanium nitride which is deposited onto the surface being coated. Because the whole process takes place in a glow discharge of argon the surface being coated is initially cleaned and the titanium nitride coating is embedded in the surface. The process is used to coat steel used in machining tools and can increase the life a drill by over an order of magnitude [14]. Figure 25 illustrates the improvement in wear properties by determining the effect of putting a titanium nitride coating onto steel drills. Figure 25a shows the X-ray diffraction pattern from the coated drill, confirming that the layer deposited is indeed TiN. The number of holes that can be drilled before the drill requires sharpening is shown in Figure 25b. It can be seen that a coated drill can drill 100 holes before requiring resharpening compared with only 10 for the uncoated drill. Suprisingly the coated drill still outperforms the uncoated drill even after drilling 100 holes and being resharpened. Clearly the TiN coating is still present, either as a result of diffusion or being embedded in the matrix during use and resharpening.

9.   SEGREGATION

Segregation, that is the enrichment of an element relative to its matrix concentration, occurs only too frequently in materials. In some cases the effects of segregation can be beneficial, as is the case when a grain boundary is strengthened by an element segregating or when small carbides form which pin dislocations and improve strength. Unfortunately segregation is only too often detrimental, as in the case of segregation of trace impurities, such as phosphorus, to grain boundaries. Elements segregate to reduce the free energy of the surface and do not always segregate to the grain boundary surfaces. For example sulphur will preferentially segregate to the surface of a titanium particle in high nickel alloys but not to grain boundaries whereas the reverse is true of phosphorus.

Grain boundary segregation is measured in terms of an enrichment ratio [9]. This enrichment ratio β is defined as the ratio between the interfacial concentration X_b in mole fractions of a monolayer, and the bulk solute mole fractions X_c.

$$\beta = \frac{X_b}{X_c} = \frac{\Gamma_b}{(\Gamma_b^o X_c)} \qquad (3)$$

where X_b and X_c are the interfacial concentration and bulk solute concentrations respectively. Γ_b^o is the Gibbs 'excess' solute at the grain boundary in mol/m ^{2} and Γ_b is the quantity of solute in mol/m ^{2}

A plot of grain boundary enrichment ratios as a function of atomic solubility X_{co} in Figure 26 shows that the interfacial activity can vary over several orders of magnitude with enrichment ratios varying from less than ten to in excess of 10000. Clearly even if the bulk concentration is low in certain instances very high levels of segregation may occur and if the segregating element is detrimental the results may be very serious. This was the case in an accident with a turbine rotor blade in Hinkley Point Power Station in 1969. Phosphorus had segregated to the grain boundary to about 0.45 of a monolayer. This so embrittled the bade that it fractured destroyed the turbine and caused in the region of £1 million of damage.

In modern alloys elements that are regarded as detrimental include, sulphur, phosphorus, tin, antimony, copper and potassium. They act to reduce the mechanical properties of the material by segregating to grain boundaries, thus weakening the bond between adjacent grains. When segregation does occur, particularly when the segregating species is a trace element in the bulk, it forms a layer only one or two atom layers in thickness. This has been demonstrated by Palmberg and Marcus [15] on many systems by depth profiling through the segregated layer and comparing the depth profile with that predicted for various thin layers (Figure 27). The segregation to a grain boundary is related to the surface segregation. Seah and Lea [16] have pubished equations to predict the enrichment at the surface and grain boundary and show that for tin in iron and segregation levels less than one monolayer the grain boundary segregation is two orders of magnitude greater than the surface segregation (Figure 28). It would thus be difficult to predict a grain boundary enrichment from a surface measurement and it is therefore necessary to expose the boundary and measure the segregation directly.

9.1   COMPOSITE MATERIALS

Increasingly use is being made of a relatively new range of materials known as composites. These consist of fibres of one material coated with a second and embedded into a matrix made of a third material. Often the fibre is made from carbon or tungsten but the matrix can be anything from a lightweight metal to a epoxy type of resin. The bond between the fibre and the matrix is very important. In some instances the fibre is diffusion bonded to the matrix with a material included to aid the bond strength. Copper is used to diffusion bond tungsten to the aluminium alloy matrix. However, this can cause problems with the copper diffusing to the grain boundaries and embrittleing the matrix. The intention with these materials, as with most new materials, is to produce a product that has either or both increased strength and reduced weight. We are very familiar with the replacement of conventional materials with carbon fibre constructions in the sporting field but it has been predicted that by 1995 one third of every aircraft frame constructed will be made from a composite material.

What are the problems associated with composites? Clearly the strength of the individual components is important and when considering a new material these are selected with great care. However, once selected, there is little one can do to improve this area. Design of the composite is then paramount. For example size, separation and orientation of the fibres is important in determining the resultant properties of the composite. These are however, design considerations and surface science will have little or no imput here. However, once the materials have been decided, the design adopted the strength of the material will depend on the bond between the various components in the matrix. By fracturing a composite and examining the resulting fracture surface one can determine where failure has occurred and analyse the failure surface for detrimental segregants etc. An example of a modern composite material which has been fractured and examined in a scanning Auger microprobe is shown in Figure 29. This is a composite material that uses long carbon fibres. These are coated with silicon carbide and the coated fibres are embedded in an aluminium alloy matrix. The three components are clearly visible in the secondary electron image. Most of the carbon fibres have broken in a brittle mode but a few have pulled out from the silicon matrix and there appears to be some pulling away of the coated fibre from the alloy matrix. Analysis of all the boundaries indicates a high level of free carbon but the carbon/silicon boundary also contains some additional metallic species. An elemental Auger map of silicon taken over the surface of the fracture face clearly shows the presence of silicon in the fibre coating and at points on the boundaries between the matrix and the fibre. Surface science will play an increasingly important role in both the construction and quality assurance of these composite materials.

10.   CONCLUSIONS

The intention of this paper has been to review the application of surface analytical techniques to the study of advanced materials. This has been done using examples from as many different areas as possible and has shown that problems do exist with many of the so called advanced materials. Many of the examples cited here are from work on the new superalloys, which contain high levels of nickel and chromium, however many of the new materials have yet to be thoroughly examined using surface analysis. This is particular true for the range of materials used in the aircraft industry. To date only the oxidation characteristics of titanium- aluminium alloys have been reported and there is even less work done on the composite materials. This is an area where surface analytical techniques can make a considerable impact. This is particular true for those with high spatial resolution such as scanning Auger and scanning SIMS. These techniques should be used to study the segregation to interfaces on fracture faces of the new alloys and composites. The strength of the alloys may be improved by altering the grain boundary segregation while the mechanical properties of the composite materials will be determined by the segregation of elements the fibre matrix interface. In short these techniques have only just begun to scratch below the surface on advanced materials problems and there is much to be done.

11. REFERENCES

1. M.P.SEAH: Surface Science, 1975, **53**, 168-212.
2. Practical Surface Analysis, 1983, Ed. D.Briggs and M.P.Seah, John Wiley and Sons.
3. G.C.ALLEN, P.M.TUCKER and R.K.WILD: Oxid. of Metals, 1979, **13**, 223.
4. H.W.WERNER: Surface and Inter. Anal., 1980, **2**, 56.
5. G.C.ALLEN, P.A.TEMPEST, J.W.TYLER and R.K.WILD: Oxid. of Metals, 1984, **21**, 187-203.
6. R.K.WILD: Corr. Sci., 1975, **14**, 575.
7. P.A.TEMPEST and R.K.WILD: Oxid. of Metals, 1985, **23**, 207-235.
8. M.ASK, J.LAUSMAA and B.KASEMO: App. Surf. Sci., 1989, **35**, 283-301.
9. M.P.SEAH: Int. Met. Rev., 1977, **222**, 262-301.
10. P.A.TEMPEST and R.K.WILD: Oxid. of Metals, 1988, **30**, 209-254.

11. M.J.BENNETT, J.A.DESPORT and P.A.LABUN: Oxid. of Metals, 1984, **22**, 291-306.

12. M.J.BENNETT, B.A.BELLAMY, C.F.KNIGHTS, N.MEADOWS and N.J.EYRE: Mat.Sci. and Eng., 1985, **69**, 359-373.

13. MOON and M.J.BENNETT: AERE Report, 1987, **R 12757**

14. J.P.COAD, P.WARRINGTON, R.B.NEWBERRY and M.H.JACOBS: Materials and Design, 1985, **VI**, 190-195.

15. P.W.PALMBERG and H.L.MARCUS: Trans ASM, 1969, **62**, 1016.

16. M.P.SEAH and C.LEA: Phil. Mag., 1975, **31**, 627.

FIGURE 1. A Modern Surface Analytical Instrument
(Courtesy V.G.Scientific).

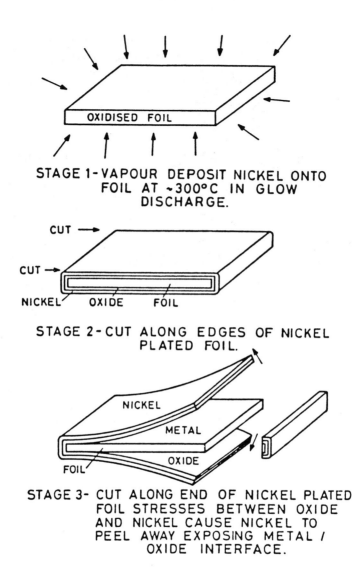

STAGE 1-VAPOUR DEPOSIT NICKEL ONTO FOIL AT ~300°C IN GLOW DISCHARGE.

STAGE 2-CUT ALONG EDGES OF NICKEL PLATED FOIL.

STAGE 3- CUT ALONG END OF NICKEL PLATED FOIL STRESSES BETWEEN OXIDE AND NICKEL CAUSE NICKEL TO PEEL AWAY EXPOSING METAL / OXIDE INTERFACE.

FIGURE 2. Technique for Exposing an Interface between a metal and oxide.

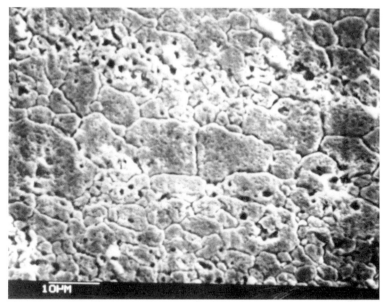

(a) Secondary Electron Image.

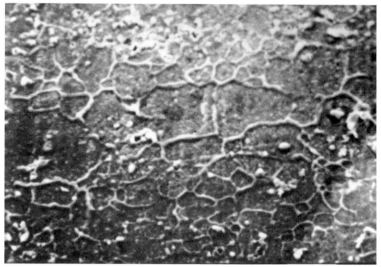

(b) Auger Silicon Map.

FIGURE 3 The metal/oxide interface on
20%Cr/25%Ni/Nb Steel.

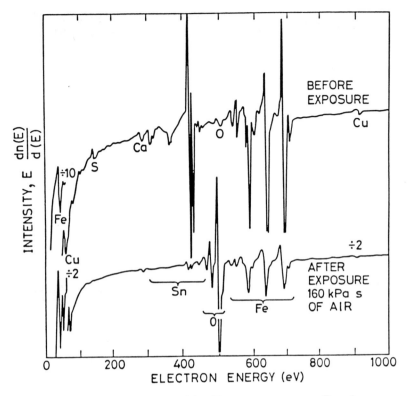

FIGURE 4. Effect of Air Exposure on Surface Segregation.

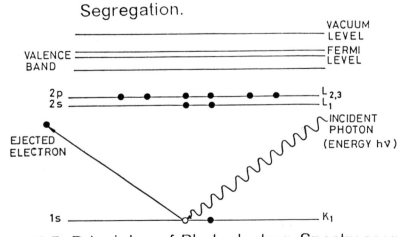

FIGURE 5. Principles of Photoelectron Spectroscopy.

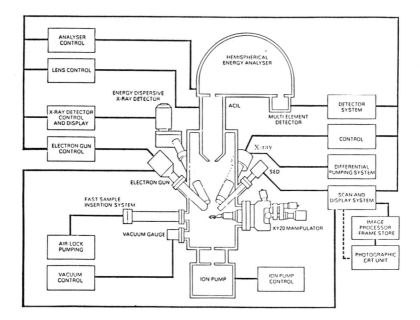

FIGURE 6. Schematic of an X-ray Photoelectron Spectrometer.

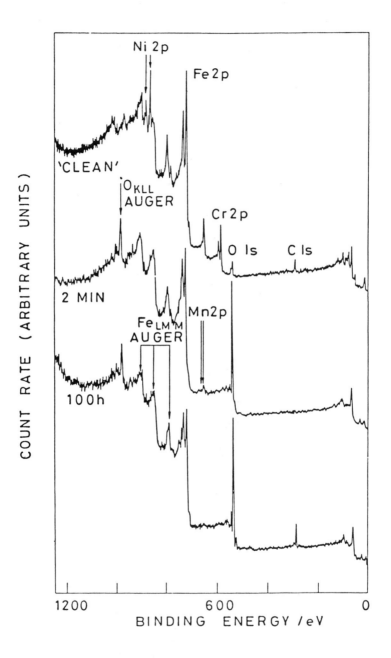

FIGURE 7. Wide Scan XPS Spectra.

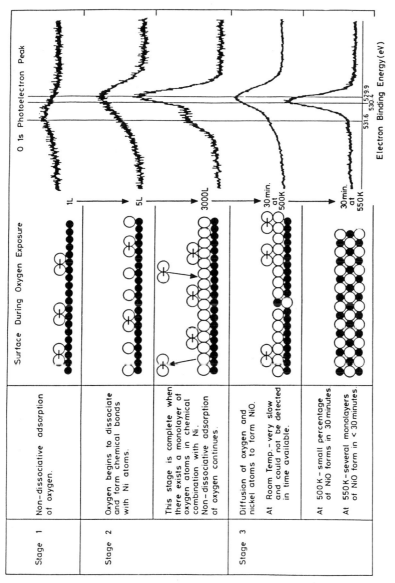

FIGURE 8. Chemical effects in XPS Spectra.

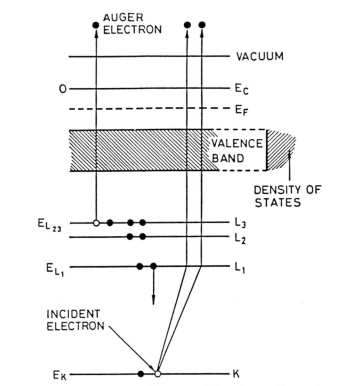

FIGURE 9. Principles of Auger Electron Spectroscopy.

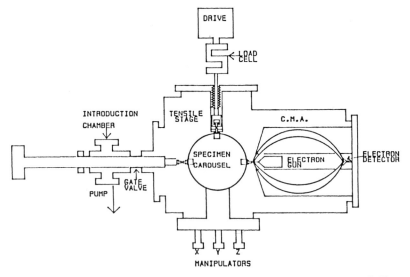

FIGURE 10. Schematic of Scanning Auger Microprobe.

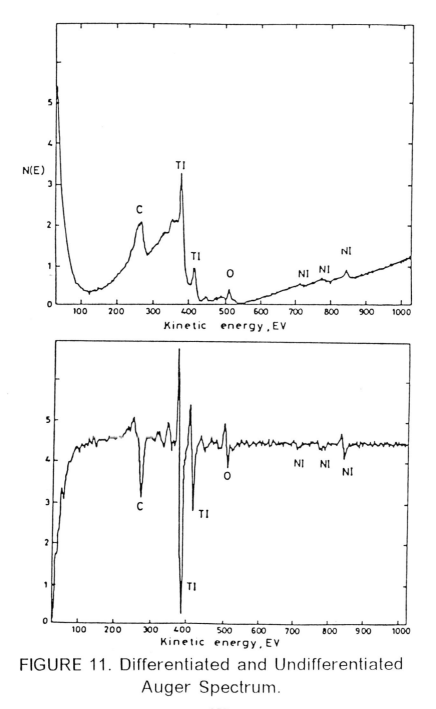

FIGURE 11. Differentiated and Undifferentiated
Auth Spectrum.

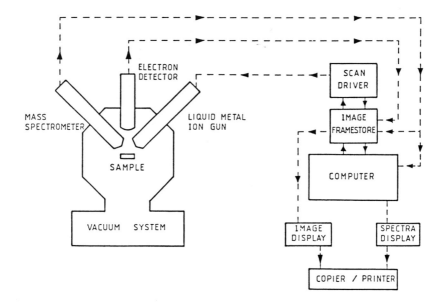

FIGURE 12. Schematic of Scanning Secondary Ion
Mass Spectrometer.

(a)

< 10 K >< 30 K cps >< 3 K

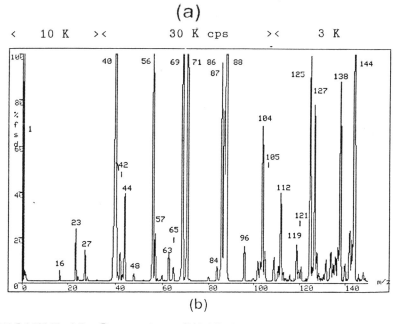

(b)

FIGURE 13. Scanning SIMS Image (a) and
        Spectrum (b) from Superconductor.

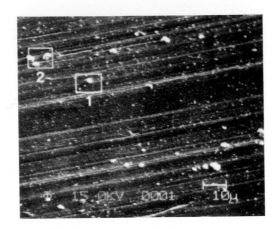

(a) Secondary Electron Image

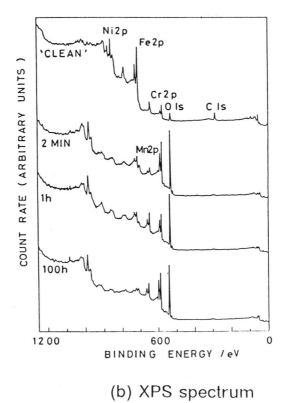

(b) XPS spectrum

280

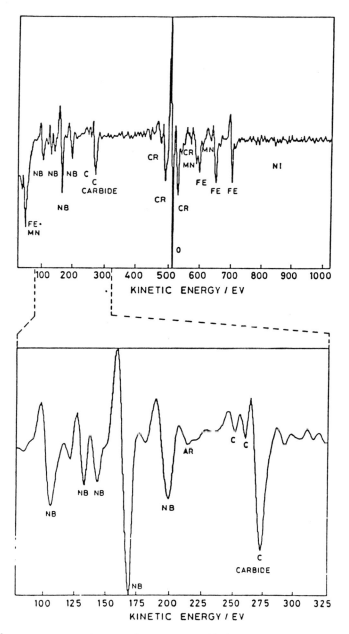

(c) Auger spectrum from particle in region 2.

FIGURE 14. Oxidation of a high chromium/nickel alloy.

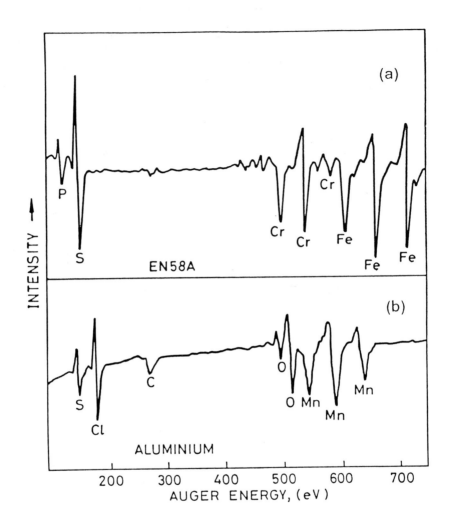

FIGURE 15. Auger spectrum from surface of heated
stainless   steel (a)
and surface of cold finger (b).

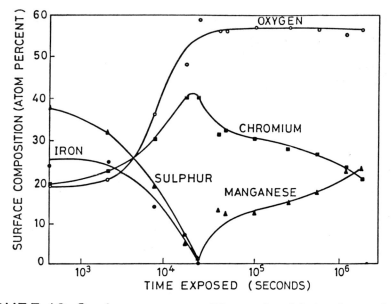

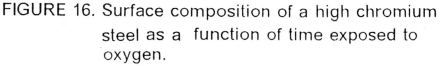

FIGURE 16. Surface composition of a high chromium
steel as a function of time exposed to
oxygen.

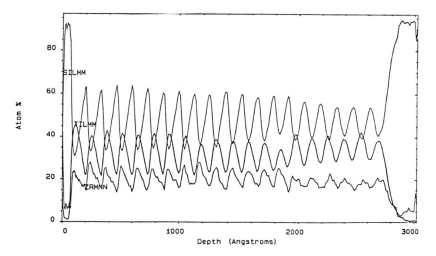

FIGURE 17. Depth profile through 40 layer sandwich
of Si/Ti-Zr. (Courtesy Kratos Analytical).

(a) Secondary Electron Image

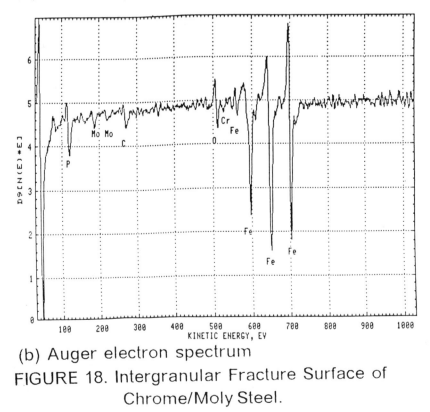

(b) Auger electron spectrum

FIGURE 18. Intergranular Fracture Surface of
Chrome/Moly Steel.

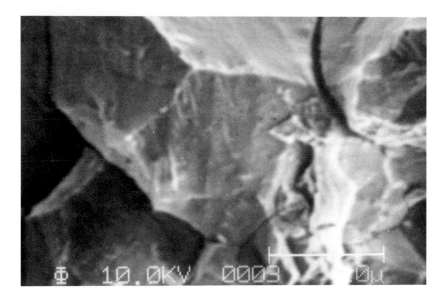

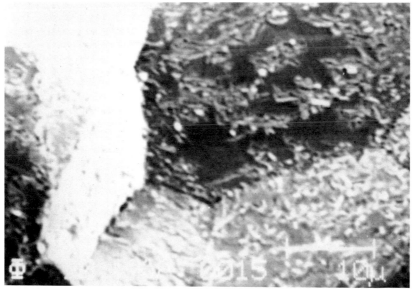

(b) Aged.

FIGURE 19. Intergranular Fracture Surfaces from
Nickel Superalloy.

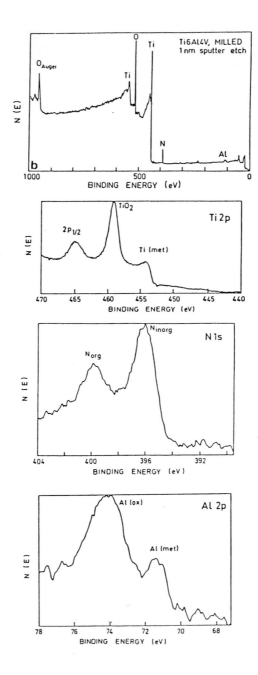

FIGURE 20. XPS Spectra from Ti6Al4V Alloy.

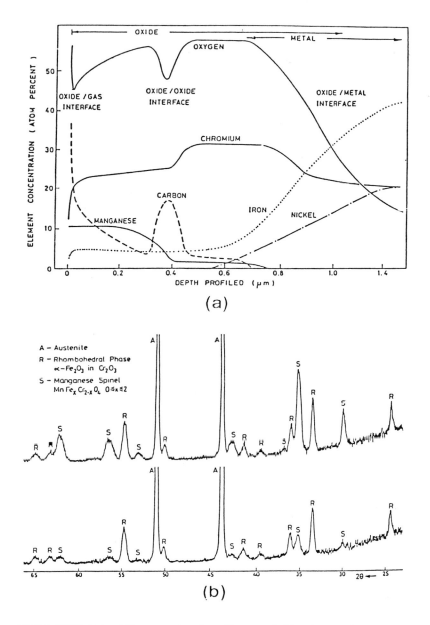

FIGURE 21. Characterisation of Surface Layers on
20%Cr/25%Ni/Nb Steel.
(a) AES Depth Profile. (b) X-Ray Diffraction.

287

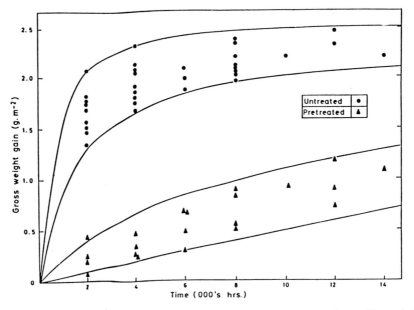

FIGURE 22. Effect of Pretreatment on the Weight
Gain of a High Nickel, Chromium
Steel at 823K in CO_2

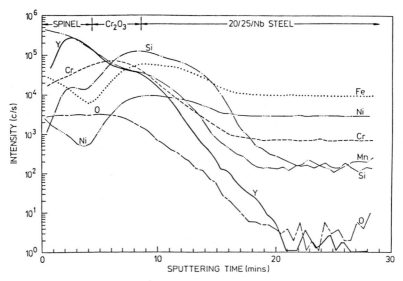

FIGURE 23. SIMS Depth Profile through Oxide
Layer on Yttrium Implanted Steel.

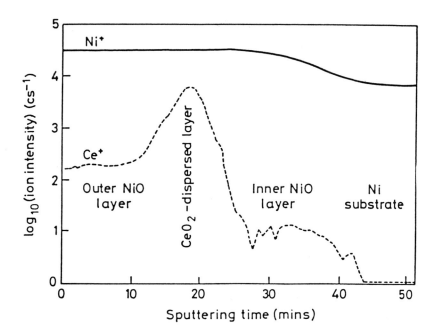

FIGURE 24. SIMS Depth Profile through Oxide
Layer on Cerium Coated Steel.

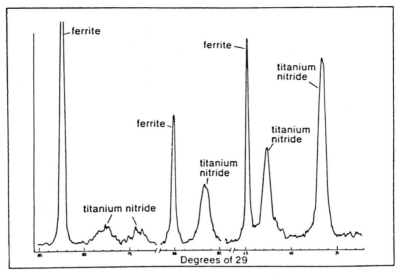

(a) X-ray Diffraction Pattern.

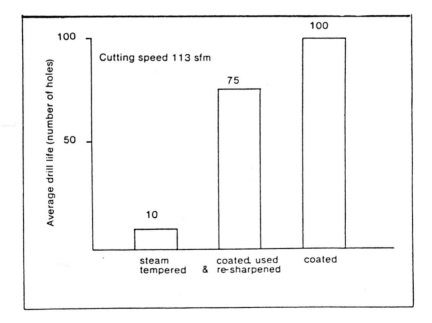

(b) Drill Life.

FIGURE 25. Sputter Ion Plating of TiN on Drill Tips.

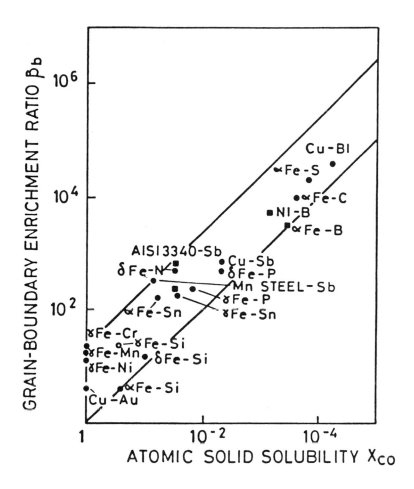

FIGURE 26. Grain Boundary Enrichment as a function of Atomic Solid Solubility.

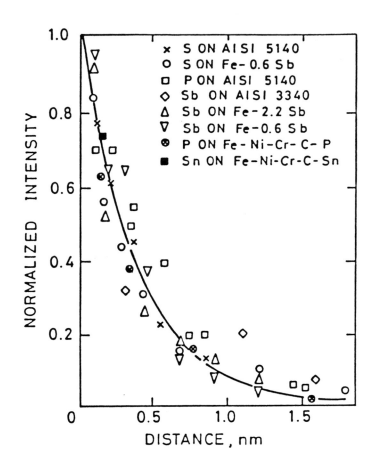

FIGURE 27. Depth Profiles through Segregated
Layers.

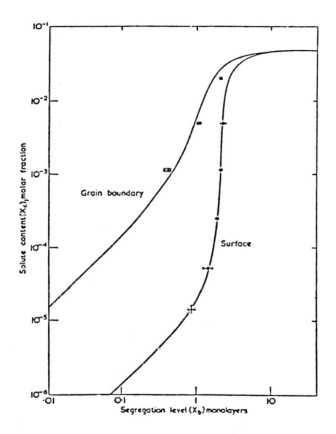

FIGURE 28. Grain Boundary and Surface
Segregation of Sulphur.

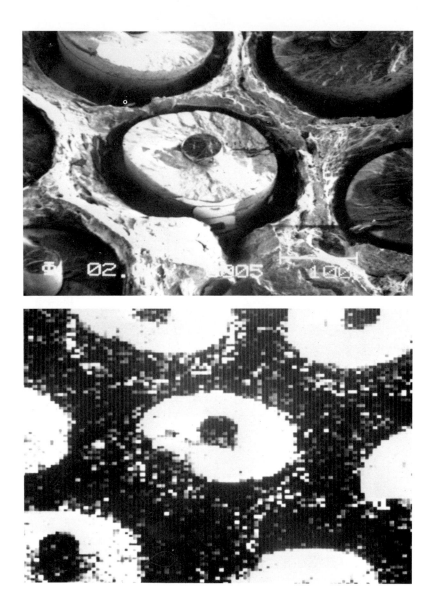

FIGURE 29. Silicon Carbide coated Carbon Fibre
in Aluminium   Matrix.